Table of Contents:

Dream Beyond The Paradigms

"Millionaires don't use Astrology Billionaires Do"
~J.P. Morgan

"Follow your bliss and the universe will open doors
where there were only walls"
~Joseph Campbell

Chapter One:

For decades great scientist and astrologers has always known how intriguing and beautiful the universe is. The entire universe is like a giant energy field, which everything is connected. All matter in the universe are composed of four major elements earth, water, fire, and air and each element rises from opposing properties like heat and cold, dry and wet. The entire universe would be considered an enigma (mystery) because know science can tell you how energy is even formed. Energy can't be created or destroyed it only changes shape. I'm sure every scientists dream would be able to travel through the universe to different dimensions at light speed. The technology was there but no one was knowledgeable of building a average car size spaceship that could travel at light speed through time but somewhere deep in the galaxies of space, a small spaceship was in motion. A young man was piloting the spaceship at light speed and could travel anywhere in space with no hesitation. This man was the great scientist Nikola Justice who was a teacher and a God. He was a God because he experienced many things in life, and a teacher not of truth because he lived his truth but of philosophy. As far as his science intelligence he mastered the electromagnetic frequencies of all the planets. His current mission was traveling to planet Saturn which is nearly impossible to enter but Nikola went to one of Saturn's many satellites (moons).

Being a scientist also made him knowledgeable of Vedic and Nadi astrology and every twelve years Saturn enters the Leo constellation. Saturn can conflict a person during this Leo transition which makes it a great time of learning and suffering, but once you are out of it you would have lived a good life and learned a valuable lesson. Saturn's powerful aura would touch Nikola showing him the Devine healings and teachings, he wanted to learn and master this power so he sat in a yoga posture and started to meditate outside of the spaceship. The spacesuit was not similar to the average spacesuit with the bubble shaped head gear and gigantic puffy like suit, the suit is made of thermal micrometeoroid garment but was a smaller and thinner design so he could move easier. This suit was an original designed by him. Out of Saturn's sixty-two satellites Nikola was on the only moon that had oxygen which happened to be named, Mimas.

Nikola meditates while closing his eyes which activates his pineal gland or some may call the third eye. Mimas had intense storms and lightening but Nikola remained in his meditation. Deep in his mind he pictured another dimension of galaxies and a beige color universe and then he reached his point where he wanted to go, the particle mind along with complete silence started the "AHH" Meditation. This was an ancient meditation used to be able to conceive. He channeled energy from the bottom of his spine to his navel to the heart and then to his brain. He stayed in this meditation for about two hours. He finally awakened from his meditation and said (Vi Veri Vivus via) meaning by the power of truth I why living having conquered the universe. He returned back to his spaceship and started to daydream in the cockpit, he discovered that daydreaming makes you more creative and smarter. He thought about the fact he has never had a love life or loved a woman or had a social life. He started the ship and set towards the planet earth to start working on a new mission, himself. The distance from

Saturn to earth would take years, but his spaceship had light speed and could get him there in hours. Indeed astrology was his passion but it drove him to also become an extraordinary inventor so he can truly experience what astrology has to offer.

He loved the beauty of the universe (heaven), which was actually beige in certain parts of it, yet the comets and galaxies looked like a living rainbow with all the colors. He also knew how the procession of the constellation and time worked in space. (Procession) is the zodiacs going backwards this means the constellations didn't go forwards but yet backwards. Observing the beauty of heaven he arrived quickly to the planet earth. The atmosphere of the earth did no damage to the small spaceship because the ship was made with a none flammable liquid metal. Nikola wanted to settle down in Paris France, upon arrival he saw It was midnight over Paris which is a beautiful city at night.

He landed on the roof of the hotel where he stayed. Nikola stood at a height of five foot nine inches with hazel brown eyes, caramel skin complexion, jet black color hair low hair cut with deep waves and a thin mustache. He was not from the planet earth but he was human, no different from any other human. The next night he arrived at the [Chateau De Chenonceau], which is one of the most beautiful castles in France on his search for crystals. That only appeared during this twelve year Leo constellation. The shallow water outside of the castle was filled with glowing stones that was a bright neon green and turquoise color. Matter of fact the moon was the same color as the stones. The crystals were not hard to find in the forest because the light from the moon showed the path to the crystals.

Nikola was told by his past ancestors always follow your heart and the light from the moon. It was truly beautiful the colors of the crystals below the tree which held them. He obtained the crystals he set out to look for and now headed back home, but he needed a place to stay the night. Nikola didn't have much cash on him but he a some dollars to check in a hotel so he went to one of the nicest suites in Paris. The hotel receptionist was very friendly to him and quite flirty. She knew from first impression he was not from the area. "Bonjour" ma Cherie Nikola said to the receptionist, and no he is not from France. They conversed for sometime and Nikola needed some work. They discussed his strengths and weakness and the receptionist had an uncle who was a scientist also.

Nikola could speak fifteen different languages so he held the conversation with the receptionist very well. The next following morning he gave the receptionist uncle a call. He was a very friendly man and told Nikola to come down to the laboratory. Now being a God from the 10th planet in the earths solar system actually served him some pretty good luck with others. On his planet engineers, scientist, and astrologist were just called teachers since they didn't have many names for the different categories but to the people on earth he would be known as a God for his profound knowledge. The 10th planet was called Nibiru just outside of the planet Pluto. Nikola arrived at the laboratory and met Dr. Borovsky. The doctor said "You must be Nikola?" Nikola said "Yes, how did you know?" The doctor replied "My niece said you were half the size of the Eiffel Tower," the irony of it was he was only 5'9 in height. The two scientists laughed for a few moments and got to business. Nikola told the doctor that he was past the current 2013 technology. He explained his skills to the doctor and knowledge but the doctor didn't understand any of the outrageous concepts.

Chapter Two:

The doctor explained to Nikola a little about the job and the year 2013 technology and that was easy for Nikola. The intelligent conversations lead the other scientists to gather around. The others introduced themselves to their newest member. Having this new knowledge the French government funded Nikola and gave him a government credit card and there he went to purchase his furniture for his new apartment that was also a gift from the French government. Nikola always thought positive and felt joy in his heart and he knew the secret to life itself. Whatever you think it manifest in reality. Whatever you think good or bad you're always right. The new apartment that was given to him brought even more joy in his heart and he even had a garage for his spaceship. The spaceship could do many things. It could transform into whatever automobile Nikola could visualize in his mind.

Six months later Nikola had rearranged one of the bedrooms in his two bedroom apartment into a laboratory so he can do his research from home. He had built a reputation of the best scientist in France and some of the other scientists were jealous of how quickly his popularity grew, but Dr. Borovsky was his good friend. Nikola was experimenting with a water experiment he had learned in Japan. He took a jar and filled it with water and wrote different things on the jars, one said I love you and the other said I hate you. He used a inferred microscope to see the results after about eight hours of freezing in a freezer. He seen the water crystals molecules took different form. The water crystals looked different just by his two different thoughts writing on the jars. The French Government knew of Nikola's genius mind, and like most governments they could benefit from him but as smart of a man as Nikola, he knew the intensions of the government was to control humanity.

The French government obtained the technology decades before Nikola's arrival to earth but they didn't have the knowledge to complete the research, of course he went along with their ideas. He mastered electromagnetic frequencies (radio waves) which is the prime factor required to control humanity but he never wanted to use this power to control anyone, because in the wrong hands this could split a planet into two parts. Nikola had to admit even though it is a great power during his time mastering it, he had so much fun doing so. He convinced himself to take sometime away from his research, due to the fact he needed some fashion. European fashion was the best fashion on Earth, so he was in the right country. He arrives to the brand new Confluence Mall in Lyon, France. This mall had four levels, and all the right stores Nikola needed. He browsed the entire mall visiting the Apple Store, European Eagle and Ralph Lauren Outlet and had purchased some formal and casual clothes. He had six bags in hand so far but not yet purchased any shoes. He had slacks, jeans, polo shirts, t shirts, business attire and he could not forget the all important underwear and socks. The government credit card had unlimited funds on it so it was at his will to stop shopping.

Nikola had one last store to browse for some shoes. His eyes glowed with excitement because he was in a store with shoes wall to wall. There were boots, casual shoes, sneakers, Top-Siders and Stacy Adams. As he was looking around for some shoes a beautiful voice from behind and said "Can I help you with anything today Sir? Are you looking for anything in particular?" Nikola dropped his bags gazing into her green eyes. She then asked "Is everything alright sir?" She started to wave her hand in front of his face because he was just standing there staring as if he was in wonderland, he looked down and her name tag read Alyssa. She was 5'8 in height with long wavy jet black hair, rosy red cheeks, a magnificent smile and the most stunning bright green eyes Nikola ever saw, she was a ravishing European woman. He whispered to himself "Your Beautiful," but apparently it was noticeable because she read his lips and said "Excuse me Sir?" He said "Yes madam, you can help me today." He needed some Top-Sider shoes, in a size nine or ten. She said "Right this way". His phone started to ring, it was from the lab but he didn't answer. He spent about thirty minutes trying on different pairs of shoes.

Alyssa told him her lunch break was about to start and her friend could help him. "Hello sir my name is Kloey, and I can help you today?" said Alyssa's co-worker. Alyssa left towards the food court inside the mall, but Nikola didn't care for the shoes any longer, his attention was only on Alyssa now. He told Kloey to load all the shoes up and he will purchase them then gave her his credit card to check him out at the register and told her to meet him at the main entrance in fifteen minutes. He also gave her a tip to get some of the other workers to carry his bag of clothes that he previously purchased. He ran out the store looking for his green eye beauty he just met bumping into person to person looking for the food court. He seen her from a distance because the ceiling of the mall was glass and the sun was beaming down shining on her dark hair.

He finally caught up to her and told the cashier at the salad bar that he would pay for Alyssa's food. Alyssa laughed and said "Aren't you the guy from the store?" He replied "Yes!" He told her he couldn't leave the mall without telling her how he felt. "You're pretty blunt aren't you?" said Alyssa, he told her green is his favorite color and her eyes reminded him of one of the earths moon phases. His powers of observation served him well, in just a few moments he notice that she had on crystal earrings that matches her crystal bracelet. A famous quote he said he learned years ago was always follow your heart and the light from the moon. Alyssa laughed and said "There is no moonlight out." They both laughed and soon got to know each other a little better. Nikola told her that laughter is always good for the human body. She knew he was not from France but Nikola was not ready to tell her the truth just yet, he only told her he was a researcher and scientist.

Her lunch break went by fast so she had to get back to work, Nikola attempted to walk her back to her job but he remembered he was suppose to meet her co workers at he entrance of the mall to load his shoes up. Alyssa enjoyed the laughter and was actually intrigued by his intellect. He was different from most guys she knew and she loved his good sense of humor so she told him to call her sometime. The two exchanged numbers, and Nikola lit up with excitement and ran to the entrance of the mall to meet the workers. The workers were upset, they all had angry looks on their faces do to the late timing of Nikola. They threw the bags in his car, Nikola said sorry guys here a tip. They didn't want the tip so they walked away saying "We have a tip for you and that's to get a watch!" Driving through the city he would have his spaceship transformed into a sporty red Italian car. He was driving fast swerving with joy, he was shaking up everything in his trunk, all that was on his mind was Alyssa.

Chapter Three:

He finally made it home and he heard something in his trunk mumbling, he thought he was hearing things. He opens his trunk and a tiger pops out. Its his best friend from planet Nibiru Chi Chi who was a walking, talking, flying tiger. She was a girl tiger about the size of a raccoon and has a raccoon tail. Chi Chi was a genetic experiment from the gods on planet Nibiru of a tiger cub and human DNA. But the raccoon tail came from her curiosity of drinking lab experiments, (curiosity killed the cats' tail). He was shocked Chi Chi been in his trunk that long period Chi Chi had her mini size spaceship and was in a cryogenic sleep for months, the banging and the bad driving Nikola was doing activated her ship to awaken her. Back on planet Nibiru animals and human walked the planet together with no fear but Nikola needed to keep Chi Chi hidden so humanity wouldn't panic seeing a talking tiger cub. He had a great idea, he would say he did a lab experiment on stray cats in the area. Alyssa texted him and said she made it home safely from work.

The two text back and forth, they were waiting for the other to call, Nikola was nervous and Alyssa was playing hard to get. Chi Chi jumped and opened to fridge and seen it was no food, Nikola said he has not went grocery shopping yet and told Chi Chi we will get something to eat soon, he was getting to know this woman he met. Nikola made plans for a dinner with Alyssa for the weekend. Nikola had O.C.D (Obsessive-Compulsive Disorder) really bad, that's an anxiety disorder where people have repeated ideas, thoughts, feelings, sensations or they feel driven to do something compulsions so basically everything had to be in its place, he was always cleaning up and wanted everything in a certain order and not to be out of place. He started playing 80s classical rock music while he was cleaning his apartment, he wanted his place spotless for Alyssa. After he was done cleaning he told Chi Chi lets go wash 'Brzee' that was the name of his spaceship. He named his spaceship after the Goddess Lakshmi who was the goddess of wealth and beauty. The spaceship could do anything the owner wishes it to.

It has its size limits it can not go bigger than its regular spaceship size but it can transform into any car that matches its size. Chi Chi was happy to tag along for this earth journey she apologized for sneaking in Nikolas trunk but she said she wanted to get away herself and she knew he wouldn't let her go. Nikola told her it is peaceful here on earth and he have more joy in his life. He kept talking about Alyssa and how much he liked her, Nikola was excited about the up coming date, Chi Chi said "Yes about time, back at home you never had a social or love life." Chi Chi was still hungry so they waited on the car wash and headed to the nearest Grocery Store. Nikola was an organic food fanatic he never ate fast foods because he knew that the beef was not 100% beef. It was about 15% beef and the meat still had the growth hormone in it and the pink slime but the sad part it causes cancer. Nikola's cell phone had a scanner on it to scan the bar codes on foods and it actually tells the history of the food. He scanned the ground beef and it told him how many cows were in the beef where the cows came from and which part of the cow.

The scanner would flash red if steroids or growth hormones were in the meats or any foods. His scanner would tell the percent of air that would be in a single bag of potato chips since most chips are majority air just to make the chips look bigger. He mainly shopped for herbs and spices because in his spaceship he had a device that would transform herbs into food. The grocery shopping went fast because they didn't need to get much, but outside the store he seen a homeless family. Nikola was against poverty but he never talked about it because of the law of attraction. It would only make things worst talking about it because that's a strong emotion and thoughts matter to the universe. He went and opened his car door and reached under his seat he had about seven hundred dollars to give the family. And on top of it he paid for a cab so the family could come eat a well cooked dinner at his place. The family consists of a mother, father two boys and a girl. Nikola had a huge heart especially for children.

They all arrive at Nikola's apartment and Chi Chi started immediately cooking, the kids laughed at Chi Chi's tail and began to pull on it, even though as a feline she hated that she knew it mad the kids happy. The father of the family was so happy that Nikola came along because through out the entire day no one would stop to help them. As pure as Nikola's heart is he told the family they could stay as long as they needed to. After eating the kids and the mother went to bed, and Nikola and the father finally got around to introducing themselves. They talked outside on the balcony so no one would wake up. Tony was the father and he said that's his wife Mohini and the boys name's was Jin and Ricky and the daughter's name was Alice. He told Nikola that Jin was adopted. Tony and his family were members of the biggest church in France but Tony started asking questions to the Pope of the church. Tony had people at his job and people around the area questioning the church also. Religion is no longer sacred to people.

Everyone questions the meaning of life and the direction that science has taking it, "It was the Age of Enlightenment," Tony said. Enlightenment does not come on the hills of the Black Plague, no that is not Enlightenment. Enlightenment comes on the hills of plenty. After Tony started this questioning, jobs were lost for the curious people, including Tony's job. Nikola said "You did well question everything because religion without science is blind, and science without religion is lame." Nikola was very open with Tony and introduced himself. He said "I am Nikola the enlightened one, I am 36,000 years old and on my planet I am a teacher but on earth I am known as a god. My destiny on earth was to plant in humanity minds outrageous concepts." His teachings were simple, to get people to realize the great intelligence within themselves. He told Tony that he mastered in Vedic Astrology and Geometry also Science was his favorite. Tony said he never met a god before but Nikola told him he has met a god, "Just go look in a mirror." We all have great powers, but the powers we share are that the soul never dies. Reincarnation plays a role in everyone's life humanity is just borrowing a body.

Tony asked him was life a test? Nikola said "Life is not a test, it's a opportunity." Nikola taught Tony a few techniques and meditations; he started off showing Tony secrets of sounds. Sound and Intelligence are synonyms. Sound frequencies are as old as space and the universe some call it mind science which also employs the use of powerful phonemes (sound frequencies) to enhance intelligence. He told Tony when he was on one of Saturn's moon he did the Ahh meditation. The Ahh meditation and Ara Kara were sounds that are ways to work with creative energies found in nature or the universe to help manifest the things you want. Dr Borovsky calls Nikola and tells him that the Senator wants him to come in. Nikola was curious, what did the French Government want with him. He told the doctor that he would be in first thing in the morning. Tony told Nikola thank you for everything and then he went to bed.

The next morning Nikola gets up and tells Tony if he needs anything Chi Chi will help him with it then he heads to the governments laboratory to see what the inquiry is about. Nikola sends a text message to Alyssa saying, "Rise and shine beautiful" and that he is excited about their first date. He told her he was busy last night and had to help a friend out. He talked to her the entire way to the lab. She was just as excited about the date tomorrow as he was. She was up early making her lunch today to take to work because she had to run some errands before work. Nikola said he was bringing her some flowers and a card to her job today, just to see her smile, he loves her smile. Alyssa told him she was going to call him by his last name because Nikola was weird. They both laughed and she started calling him Justice from then on. He arrived at his job and told Alyssa "Until tonight my dear, I bid you a farewell".

Chapter four:

Doctor Borovsky met him at the entrance and they went to the business room. He never met these new faces but point in fact they were the leaders of the world. The American, European, Asian and Russian leaders of the world. The Senator told Nikola to have a seat. What they needed was his knowledge to be able to transmit human thoughts. The Senator said we killed over three dozen people trying to experiment on the human brain, but we don't have the right frequencies. Nikola was the only one on planet that knew that energy of about one-hundred fifty hertz of what the earth naturally produces can trigger the human brain. The Asian government told him to develop a pill that couldn't be digested for seven years and to include it in the drinking water or coffee to spread over the globe fast. Nikola already knew about the computer chips the government has plans with to release soon to control human behavior, but they needed his knowledge to protect human thoughts. The pill would go into the human stomach and would be like chewing gum and not digest for seven years, but the pill would be a receiver and inside the pill would be radio waves at about one-hundred fiftieth to transmit human thoughts and with internet on cellular devices and Wi-Fi it would transmit back to the government under ground facilities.

The government wanted nothing hidden, not dressing rooms, department stores, street lights, social networks, home clocks, and even door handles had cameras, it was no privacy. Nikola asked how much would this project cost. They told him thirteen billion dollars. He said with that you could clear world poverty for thirty plus years. They said "This is not the time for compassion Mr. Justice." Nikola told them to give him thirteen billion years to think on it and he would get back with them. The senator said he thought Nikola wouldn't agree and a team of soldiers are on the way to take back property of the apartment and the merchandise that he purchased so Nikola called him an Indian giver. They knew Nikola didn't have any family that's why they couldn't intimidate him with any of their scare tactics. Nikola started to walk out the door with confidence because to him losing his apartment was not a big deal. Nikola asked "Is that all you guys needed today?" They said "Yes that's all for now, unless you want to talk about Alyssa." Nikola stopped in his tracks and took a serious look on his face.

The Russian leader said Nikola knows too much and they demanded he wouldn't leave. The government had his phone tapped so they knew about his passion for Alyssa. They asked him how was their relationship and where was he taking her for the date tomorrow. Nikola knew it was trouble now and they ordered for her capture and had soldiers on the way to her job. Nikola said "DO NOT TOUCH HER!" They wouldn't let Nikola leave because he knew too much and they needed his knowledge to take over humanity. The Senator called three soldiers to take Nikola away and they will deal with him later but Nikola was not going out without his love. Alyssa meant too much to him and in a short time to lose her like this. He fought the soldiers and Doctor Borovsky opened the door for him to escape so Nikola ran outside to his spaceship to try to rescue Alyssa and Chi Chi, but he soon found himself surrounded outside. The Senator ordered his capture and they had tanks, helicopters and jets outside along side about one hundred armed soldiers. Nikola was a distance from his spaceship, but luckily his thoughts would activate the ship.

He needed to be two places because with his escape they would kill Chi Chi and Tony's family but he needed to save Alyssa. He couldn't use light speed on the earth because it would rip a hole in time and affect the earth's magnetic poles. He told his ship to fly to him and he jump highly inside it. They started firing but his ship was bullet proof, he had no way of calling Chi Chi to warn them, but he went with his heart he had to get Alyssa (the power of love). Back at his apartment Chi Chi happened to be on the balcony just like a nosey cat. She noticed the soldiers coming up her senses told her it must be trouble. Chi Chi told everyone "Out the door, quickly hurry." Chi Chi grabbed the bag of crystals but she was clumsy and dropped a few. Jin helped her picked them up, the kids said pretty rocks and quickly ran out the door, Chi Chi got everyone down the staircase safety and told them to wait behind the garage.

She went back up the staircase but the soldiers had raided the apartment. They were ordered to look for any formulas or documentation possibly giving them the technology they needed. They came up short and the Senator said "Tear the place apart and if there is nothing head over to the mall and help with the capture of Alyssa and if you see Nikola shoot him on site." They said "Yes sir!" Chi Chi over heard the conversation and was not happy. She asked the soldiers could she help them find what they were looking for. The soldiers laughed and thought she was a toy. Chi Chi said she was pretty much real and she can show them. She flew across the room punching one soldier and kicking the other. They started to take fire but Chi Chi was to fast for bullets. She slapped the other soldier with her tail and the last soldier tried to run but she tackled him. Chi Chi was strong she was actually a thirty year old woman in a tiger cubs body. So she had strength and speed of a human, tiger and raccoon. She needed directions to the mall so she asked one of the injured soldiers but he said he didn't know. She pulled her claws out to rearrange his face, then he told her the directions to the mall. She could sense the fear in his eyes so she knew he was telling the truth. She knocked him out with her tail and went over to inform Tony.

Chi Chi told Tony and his family to go to the hotel about four blocks from the apartment, where they could stay. Chi Chi had to get to the mall as fast as she could. She leaped as high as she could and started to fly into the air, heading towards the mall. Nikola was flying as fast as he could to get to Alyssa; he made attempts calling her cell phone over and over and got no answer. He started text messaging her, and still no reply. He called again and it was going to her voicemail. He hung up and turned corner after corner until the helicopters caught up with him and began to fire. Alyssa calls at the same time Nikola was calling that explains the voicemail, but during the impact of the helicopters Nikola drops the phone. The helicopters radioed in the cyclone jets that had missiles.

Even though Nikola spaceship was bullet proof missiles would do some damage. Alyssa kept saying Hello, she heard him yelling in the background but he was trying to keep the jets off him. She could barely hear him and said she was five minutes from her job and that she was on the phone with her boss telling him she was running twenty minutes late that was the reason she missed Nikola's calls. Nikola pulls up higher to get a chance to reach his phone. He grabs his phone and tell Alyssa not to go to work. She replied Justice I have bills and I need my job, besides I am already late. Nikola told her that government military was trying to capture her. She said what for? He told her to trust him. Alyssa arrives at the mall the soldiers had a image of her and seen her. She pulls her car into the parking garage. She seen them approaching her so she took Nikola's advice and kept driving up the ramp in the parking garage.

Chapter five:

The soldiers started shooting at her tires. Alyssa started screaming and exited the car. Nikola told her to run and find safety. He was on his way but he couldn't approach the area with the jets attacking him. He went zooming around every corner building to building making one of the helicopters hit a building. He stayed on the phone with Alyssa listening to her screams as the soldiers continued to chase her up each level in the parking garage. She asked Justice where was he, he said he is on the way. Nikola had to get the jets off his trail because it would mean more danger for Alyssa with the rapid firing. Alyssa ran to the top level of the garage. The thirteenth level was the highest and the soldiers were right behind her. One soldier approached her and grabbed her hand and said come with us but she turned and punched him in the face. The soldier hit her in the back and her phone feel out her pocket. Nikola heard all the activity so he started pushing his spaceship as fast as it could go.

The soldier lip was bleeding from her punch he immediately slapped her to the floor. The other soldiers said "The senator wants her alive." The soldier that hit her said "I won't kill her, but I am going knock her unconscious with my rifle." The other soldiers didn't want to see it so they turned their heads and pretend that they didn't know what he was about to do but because he was upset he was bleeding by a woman they knew he wasn't going to let that slide. He raised his rifle about to strike her, but Chi Chi arrived and kicked the soldier directly in his nose. Chi Chi said "Never hit a lady." Chi Chi asked Alyssa has she talked to Nikola, she replied "Oh My Goodness, where is my phone?" he was still on the phone. Alyssa gave Chi Chi the cell phone and Chi Chi said "Whatever plan you have Nikola you need to hurry. The soldiers started to rush Chi Chi, she threw the phone back to Alyssa and fought the soldiers.

Nikola seen the action on the top of the garage but he couldn't just land on the top floor and risk his friends getting shot. The jets still remained behind him, but he flew directly into the entrance of the parking garage. The cyclone jets followed behind him and were too big to enter the garage. The impact from the cyclone jets caused the garage to start falling apart, he knew if he didn't get to them fast enough they would be buried alive. Nikola changed his spaceship into a sporty Italian coupe so he can make it to the top faster, because of the falling debris he couldn't fly. He drove fast from the first floor trying to reach the top with the floor beneath him collapsing. Nikola was on the 8th floor, he knew he was not going to make it so he told Alyssa to jump; she looked over the edge of the garage but didn't see Nikola. She said "I am scared of heights."

The soldiers saw they couldn't take down Chi Chi, so they decided to open fire. Nikola said "JUMP!" Chi Chi pushed Alyssa over the edge and she went behind her at the same time Nikola drive thought the safety rails on the 11th floor. Chi Chi slowed Alyssa's fall down by holding her arm and trying to fly but she was too heavy. Nikola was just two levels down so the two ladies landing right in the car. The soldiers started shooting and Nikola told them to hold on, he transformed the car back into the spaceship and flew away. The soldiers went down with the entire garage destroying all the jets that were following. At that point the three had to get out of site. Nikola dove directly into the ocean east from the mall. The spaceship could with stand the depths of the ocean it was no different than traveling in space. His spaceship had oxygen up to 9 hours and recharged whenever there was oxygen.

Chi Chi said "I'm taking a cat nap this was too much for a cub in one day." Alyssa was observing the beauties of the ocean, she never seen underwater cities and so many different sea creatures. Nikola said "I hope you didn't mind having our date this early," She gave him a mean look. He said he was kidding, Alyssa said "I broke two nails, was shot at, and was pushed over the edge of a garage by a cat." Nikola said "That's my best friend Chi Chi. It's a long story I will explain later." Nikola explained everything to Alyssa and she understood why the government was after her. She looked at him and said "You don't look like a god." He replied "Yes I hear that a lot." He said every human is a god. He told her he was 36,000 years old but in earth years he was 32. Alyssa liked Chi Chi, besides the fact she saved her life. He saw the mark on her arm and asked was that recent? She said it was her birth mark. Nikola told her that birth marks are really from your previous life.

Nikola asked Chi Chi where was Tony? "They are at the hotel a few blocks from your apartment" replied Chi Chi. Alyssa was tired after all the running, she rested her head on Nikola's arm and he tilted his head on top of her head. After a few hours under the ocean, they came to surface, Nikola needed to check on Tony before they departed. They arrived at the hotel, Nikola woke the girls up and they all went inside. Tony happened to be at the entrance working, he had an old friend that gave him a job and provided a room for Tony and his family with no charge. Nikola was happy things worked out for the family. He told Tony they were leaving out to America. Tony said "If you ever need anything I will be here at the hotel or move to the new apartments they are building across the street." Everyone said their goodbyes and they went back to the car.

Inside the car Alyssa kissed Nikola, he was speechless. She said "That's for saving my life." They both started kissing again, and Chi Chi came between them and said "Can we get going, I'M HUNGRY!" They all laughed and headed toward the United States of America. Alyssa was excited about the trip to American because the military wanted them dead in France. Nikola kept his money inside his spaceship so he had enough to leave the country.

Chapter Six:

Three weeks later they ended up in California. Alyssa always heard California was the best state to live in America so Nikola gave her what she wanted. They built a multi million dollar home that consisted of four levels, two elevators eight bedrooms. Nikola had started teaching Vedic astrology in America. America didn't have Vedic Astrology that was only in the east. He learned Vedic astrology in India years ago. Vedic Astrology is really a super science similar to any Science, Chemistry or Biology. They were off the east coast of Los Angelas; Alyssa loved the palm trees and the beautiful weather there. She had started her on clothing line because the European dressing was different from America. The poverty rate was higher in America than in France, but Nikola didn't join groups to talk about it he just ignore the problem. Giving his attention to poverty would only develop the situation to be even worse. He lives his life by the law of attraction, what you think is what you become meaning he would just create more poverty.

He had research groups and labs over California, teaching Vedic Astrology and soon it reached more states. Over a period of four months it reached forty-five states. Vedic Astrology can tell you if you're going to be wealthy or not. And if not then you can do something about it with the good and bad planets. If your planets were lined up correctly then you wouldn't have to do anything, money would come knocking at your door. This profound knowledge would change humanity into the golden age. Nikola inspired the homeless not just gave charity. They needed inspiration and coaching and he got his knowledge around the country. For the first time 95% of the people made miracles come true. No more poverty, stress, low income jobs. People actually believe in themselves and lived an abundant life.

The United States government started raising taxes, but they couldn't control the people anymore. No one lived in fear anymore this change in humanity got around the globe fast and also to the French Government. The world leaders contacted the senator back in France. They needed a plan to enslave humanity and put their minds back in the bubble. They insisted on Lord Uran but the senator from France didn't want to converse with Lord Uran from the planet Uranus because he would be pretty upset with this failure. Lord Uran has giving government's wealth, technology, and formulas for man made diseases. Humanity was actually living the life, stress and illness free. Nikola taught them that the human immune system can heal the body and cure you from and disease without medication or treatments. It's all on how you think. After an hour the senator had no choice but to use the military satellites to contact Lord Uran. He was the God of planet Uranus.

The French senator used the computer to connect with the satellites in space. Lord Uran lived deep, deep in his ice fortress. Lord Uran wanted control over the earth, so years ago he came up with sinister ways to enslave the human race. He created diseases and got the French senator to spread it over the planet earth. It was his idea to keep humanity thinking inside the box and to keep the human race on their knees. The senator was afraid to talk to Lord Uran, because he has failed him. Lord Uran already knew about the failure of the senator. The senator asked him, "How did you know?" Lord Uran said "The planetary Gods can connect with other planets through satellites. The universe is all connected all energy." Lord Uran says "Before your time senator, the planet earth moon was not always there. Who do you think brought it there?"

Years ago Lord Uran took one of planet Saturn's moons, since Saturn has the most satellites. He towed the moon to planet earth to brainwash humanity. The moon is hallowed and was also made inside out. The American space researchers tried to bomb the moon in the 1970's and it rung like a bell. Lord Uran asked the senator "Did you not learn anything? No planet moon is as big as the planet. The moon is 2,189 diameters, and also bigger than planet Pluto. Satellites are never big as the planet it belongs to. Therefore the planet earth's moon really belongs to Saturn." Lord Uran told the senator that in thirteen years when the rings return to Uranus he will invade the earth and do away with Nikola Justice.

"Why thirteen years my Lord?" said the senator. "I cursed myself ages ago taking one of Saturn's satellites. So the curse is Saturn drains me of my rings and leaves me in a deep sleep for years. But when thirteen years passes our rings will return and I will come. I know this Nikola is in as you call it California. We will illuminate him and enslave the human race. Spread over the media senator of an alien attack over California in thirteen years. The more lies we spread through journalist and the media the more terrified humanity will be. Nikola knows me well. We met on his planet Nibiru years ago." The senator asked Lord Uran, "Your appearance on the earth will it not do damage to the magnetic poles?" Uranus is 7 times the earth's size. With Lord Uran himself even coming in his spaceships he holds the power of the entire planet. Lord Uran said "It won't take long to kill Nikola. We will be gone off that ratchet planet. We will see you in thirteen years senator."

The computer screen went off and the senator had to start a panic in a few years. Doctor Borovsky heard everything and had to locate Nikola and tell him the news. Nikola got rid of his old cell phone because of the French government, so the doctor was fishing in the dark but he didn't give up. Back in California Nikola splashes water in Alyssa's face in their pool. Alyssa starts to spray him with the hose. Chi Chi was on the grill cooking out with her chef hat on. The two love birds go underwater, they swim with each other side by side then they go up side down and started to kiss. Alyssa told Nikola she loves him more. Nikola tells her he loves her more. She asked how much? He said a sideways 8 (forever). Alyssa never knew life could be this beautiful and abundant, not just the wealth as far as in money but they were happy and joyfully in love. He said he will never leave her side. She replied she knows. Chi Chi said "Food is ready you two, better come get some before I eat it all."

The family started to eat but Nikola started to clear his throat to get Chi Chi attention. Chi Chi left the dinner table to let them two be alone for once. Nikola said he is happy he went to the mall that day in France. Alyssa said she is happy also. The doorbell rings. Alyssa said she will going to answer the door. She opened the door but seen no one. It was a large size present box. She smiled and opened it at the door. It was a beautiful diamond ring with a card. She read the card out loud and it said will you marry............. and Nikola was behind her on his knee and finished her sentence with Me. She started to cry and answered YES YES YES YES.

Alyssa told Chi Chi you sneaky cat. Chi Chi went out the back door her and Nikola had set that stunt up to surprise Alyssa. Chi Chi made plans for the up coming wedding. Alyssa said she wanted to go out and get some fresh air. Nikola and her went to go shower and get dressed. Nikola was dressed first he wore a grey blazer and blue jeans with a pink and grey collar polo shirt. Alyssa was still upstairs putting on her earrings. She started walking down the stairs and Nikola stopped in his tracks looking at her beauty. Alyssa wore a grey mid sleeve dress with grey suede high heels her silver and pink diamond earrings her pink and grey scarf and her grey and white designer purse. Alyssa was at the door ready but Nikola was still in shock. Alyssa grabbed the keys and said she is driving. They purchased several cars after moving to America. Alyssa was driving one of her cars.

She had a Japanese sports car that not street legal. But they only made two models of this beauty machine. Nikola was scared of her driving because they drive different in France versus the United States. The weather was perfect outside and the couple decided not to go inside any where. He told her to head toward the beach. Nikola owned a few of the beach houses so he knew where the best beach was to enjoy the sunset. The couple walked holding hands kissing one another. They kicked off their shoes left them in the car and headed towards the water. Alyssa asked, "What is that green and white object in the water?" As they got closer she asked was this a bed? Nikola said yes. She smiled so big and said she loved it. It was a romantic outdoor canopy bed Nikola had ordered from Paris.

The outdoor bed was boat shaped floating bed, with totara wood which is water resistant and covered in Italian fabric. She was so excited they both got in and started to float. They got close to one another and look at the beautiful green phase of the moon. They started to kiss. Nikola put his hand on Alyssa's cheek, and looked her in the eyes and said he never understood the emotion of love until now. He always loved humanity but never felt that his every reason to breathe was for her. Nikola made the boat lift towards the air out of the water. Alyssa started to laugh and said "No Way!" She told Nikola he is full of surprises. They went up towards the direction of the moon.

Nikola started kissing on her neck. He took his shirt off and kissed Alyssa's hand. He took her dress off and said he wants to be a part of her life forever. They both laid back on the soft mattress of the bed. They started making love and looked each other in the face. Nikola put his head beside her head and Alyssa was looking at the moon and enjoying her pleasing. The moon was so bright it was though they was in front of it but was still on the same planet. Alyssa was scared of heights but she was with her lover, nothing beats the power of love. She was so into Nikola passionate love making she didn't even car about the airplane in the sky. She flipped Nikola over and got on top of him. Nikola started pulling her hair backwards while she getting more pleasure on top of him. Nikola had set auto pilot and navigation on the floating boat so he knew the course. Nikola lifted her up holding her in front of him and still inside her. It was plenty of room on the king size mattress. He turned her around with her back towards him and he kissed her ears and started making love to her from the back. She was screaming of joy but in the sky no one could hear her. Three hours went by they landed back on the water. Nikola said this was a pre honey moon before the wedding, they both laughed.

Chapter Seven:

The next morning Nikola was sleep and Alyssa went downstairs for a cup of coffee. Chi Chi was awake and asked "Did you two have fun last night?" Both the ladies started to laugh. The phone rung and Alyssa went to answer it. Alyssa had clients ready to start her new clothing line called GGG (girl go green). She went to go get ready and had to head out. Nikola got up and told Alyssa he loved her and he was in a hurry because he had a seminar to do at the research lab. Nikola and Chi Chi were on their way to the lab. Nikola reached the lab. It was over six-hundred plus teachers and other scientist there at the research lab. Nikola started off saying "WE must pursue knowledge without any interference of our addictions, and if we can do that we will manifest knowledge in reality and our bodies will experience new ways, new chemistries, new holograms, new else where if thoughts beyond our wildest dreams. The only way we will be great to ourselves is not what we do our bodies, but what we do to our minds."

Nikola started teaching more and new techniques and meditations. Nikola talked about passion and how passion is a great asset. If you have passion you don't need anything else. Whether you have money or not, but if you have pure passion it will do everything itself. He started talking on love lecture now. Alyssa changed his entire life, he never experienced love before. Sometimes, people who are together have a great age difference between them. They are together NOT because one is sexually attractive and one has other assets, but they are together because their souls long to be together and it matters not what the form, this is love that had moved here. The energy starts to move in your mind. As long it lives in the sexual area we are conditional. They all loved his speech and his teachings got all over the nation. The crowd seeing Chi Chi for the first time made them all true believers of pursuing more knowledge. Alyssa had met some of her associates.

After her meeting she wanted to go and purchase a guitar but the guitars was different in American versus in Paris, she found one that had her interest. She purchased it because she loves to music. She never got a chance to purchase one since her stay in America. She had plans to do something special for Nikola because he made her realize why none of her past relationships ever worked out. She made it home before Nikola did, so she started playing a beautiful melody. Nikola and Chi Chi were on the way home and came up with ideas for the upcoming wedding but Nikola wanted to do something no one has ever did before. Chi Chi asked "What are you going to do?" Nikola wanted to get married in space, Chi Chi said "Space? Nikola said "Yes, A third person wouldn't fit in his spaceship only two adults and a small animal," so Nikola had to have a wedding on earth but he wanted to do something outrageous.

They finally arrived at home and heard singing. Chi Chi asked was that the TV? "No, Nikola said, "That's my sweetheart." They went out back, she was on the back patio. She was quite nervous because she didn't want anyone to hear her, Nikola said "Baby that was beautiful." The big day was here, the big wedding day. Chi Chi had set up all the decorations and helped Alyssa pick out a dress. It was quite a large wedding over four-hundred people came to see the lovely newly weds. Nikola flew the spaceship to the wedding because they had planed on going to outer space afterwards. The wedding was a beautiful ceremony that even had Chi Chi shedding a tear. The bride was in all white with pink and blue lace, the groom was in an all grey suit with a blue under vest and a pink and blue tie. The bridesmaids were all dressed in all pink dresses laced in blue and all of the groomsmen were in all grey tuxedos with pink under vest and blue ties. They said goodbye to everyone that came to the wedding and took flight with Chi Chi. Alyssa asked did they need spacesuits? Nikola said "No because the spaceship had plenty of oxygen and they were going to Mimus, one of Saturn's satellites."

She was so excited but it was nothing new for Chi Chi to see because her and Nikola traveled all the time ages ago. Alyssa seen the earth as they got further and further away. Nikola got to use warp speed because the universe time didn't exist. Alyssa said "This is amazing." Nikola got there in a few hours, they landing and he showed her that there was no need for air. Alyssa trusted him with her heart and they did just that, traveled the universe together. She was nervous looking at planet Saturn gigantic rings. Nikola hopped out and said "Will you take my hand madam?" He helped her out and walked around the rocky moon. Alyssa never imagined that she would be this close to a planet other than earth, she noticed how beautiful Saturn's rings really were. While they were taking there romantic stroll, Chi Chi set up a beautiful white marble table with a bottle of Henri Jayer Richebourg Grand Cru, Cote de Nuits, France (wine) in a bucket of ice and two crystal wine glasses engraved with their initials. A white rose was the last thing Chi Chi had to put on the table so everything could be perfect, as she placed the rose down she saw Nikola and Alyssa walking back so she disappeared back inside the ship.

Alyssa said "This is unbelievable." Alyssa stopped before she took a sip of wine. She just thought about her menstrual cycle has not come on. It has been three weeks and she was late. Nikola said "Well looks like we have a baby Justice added to the family." She asked "What if I get fat?" He said he will still love her. Alyssa said "What about when I get old in age." Nikola said "My heart will still beat for you." Alyssa knew in her mind that she was human and Nikola was from another planet so he didn't age. Alyssa had thoughts in her mind and she wished that she could become immortal so she could be with Nikola forever. Little dose she know that her wish will come true, just not in the way she expect it. Nikola educated her well on astrology and how the planets work.

They did their play marriage on the moon and headed back to the spaceship. Alyssa looked out the window on their way back. The universe was more beautiful to her than the underwater city they discovered. Seven months passed by and it was time to go see what the Justice family was having at the doctor. They walked out the door and the phone had rung, it was doctor Borovsky. He finally got Nikola number after years of looking for him. He kept calling then finally left a voicemail telling him about Lord Uran attack in thirteen years but after saying Lord Uran's name a high frequency went through the phone crushing everything in his head. The great doctor fell to the floor. The senator had the labs phones and every scientist cellular phone tapped so if you say Lord Uran's name you would die a moment after. Alyssa told Nikola to drive faster, her emotions had her snappy. The family went in and was super exited to find out what is the new addition. Alyssa wanted a daughter and Nikola said it didn't matter he just wanted to be a part of her.

Chapter Eight:

The doctor came back with the results. Everyone was nervous but with excitement. The doctor said "It's a boy." Everyone started jumping with joy. They all went afterwards getting boy clothes and they needed to set up a nursery. They went to take family pictures. Nikola took a picture kissing her stomach then Chi Chi took pictures with Alyssa. The last picture included all three of them. On the way home they were still deciding on boy names. After a few hours they finally agreed on Fabian. Fabian Justice was his soon to be name. Fabian was one of Nikolas old friends on his home planet. They finally arrive home and Alyssa goes upstairs to put something more comfortable on to relax. Nikola tells Chi Chi "I never imagined loving someone this much." Chi Chi replied "Yes you two are a match made in heaven literally," (because they traveled in space together). Nikola heads to the kitchen to get some water. He checks the voicemail because he notices the red light flashing on the telephone. The message starts and he tells Chi Chi its Dr Borovsky. He was excited to hear from an old friend, until he heard the reason of the doctor's call. Nikola drops the phone.

Chi Chi asked "What is going on?" Nikola stood in shock not even answering Chi Chi's question. Chi Chi listens to the voicemail. "I Will fight for my family," Nikola said. Chi Chi asked "What are we going to do?" Nikola said "We can not tell Alyssa right now. It is not healthy for the baby." Alyssa walks down the stairs excited about the pictures they all took. She notices a different look on Nikola's face. She asks was everything ok? Nikola says "Baby yes." She grabs Nikola and says "I know my husband, and something is bothering you." She tells Nikola "Listen, we traveled the universe together, we got married, we flew on a canopy bed, we have a baby on the way together and you saved me from a falling parking garage. It is nothing we should hide from another." Nikola finally breaks down and tell her that a God name Uran would be coming to the earth in thirteen years. Alyssa says "Ok and? So what I married a God." Nikola smiles and says "True baby, you married me yes but I am not out to do harm to the people of the planet earth. Lord Uran is coming here to California in thirteen years to kill me." Alyssa started breathing fast and takes a seat. Nikola said "This is exactly why I didn't want to tell you."

Alyssa says she is ok and she said she will fight. After Nikola seen Alyssa love for him he knew he couldn't give up on his family. Chi Chi said she is in no matter what the cost. Nikola sat beside Alyssa and told her about Lord Uran. "Lord Uran was before my time. He was a scientist also on our home planet but he is 95,000 years old before my time. He tried to enslave the gods on Nibiru but our ancestors put a stop to it. It was a war and they sent Lord Uran to planet Mars. After defeating him he was imprisoned forever on Mars, but after that he escaped and left a face buried in Mars core. He set out over the Universe stealing different planet's moons. The outside of the moons is just like a onion, to get to the center you have to keep pilling. Every moon in the Universe is different and has a power source in the inside. Back in the 1970's the American and Russian governments flew to the earth's moon to see what new research they could discover. American scientist tried to blow up the moon but the moon rung like a bell. The moon was actually hollow. The inside is the beauty of the moon, it is made inside out. Lord Uran towed the the earth's current moon to the earth decades ago. The planet with the most satellites is Saturn so Lord Uran took one of Saturn's moons to attempt to brainwash humanity. The moon puts off different vibes to humans. Humans are made up of seventy or eighty percent water, so ask yourself, if the moon holds the tides (rivers, oceans, lakes) of the earth, then what is the moon doing to humans. Lord Uran couldn't enslave our home planet so he decided to come to earth. Giving governments new technology he would rule the earth and the French government knows they can not stop me so they called him to do the work. It's not the fight I'm just worried about it's the point that he holds every energy source that Uranus contains. Meaning planet Uranus is seven times bigger than the planet earth so if Lord Uran was to come into the earth atmosphere the north poles would flip to south and south to north. Causing earthquakes, tornadoes, hurricanes, and many would parish. Lord Uran would come and try to kill me and leave before the planet is completely destroyed. With him coming to the earth and if he succeeds then humanity would go back in a dark age. The French government would broadcast my death and people would panic. I stopped poverty and I gave everyone the secret of the universe, the law of attraction. I will fight for my family and the future of my family. I came to the planet earth to obtain the crystals. The crystals do many different things but the sun crystal would protect us with the sun's energy. If we place the sun crystal on top of the great pyramid in Egypt then the sun's energy would channel to every pyramid over the planet. Its pyramid's all over the earth even under water. So we have the sun crystal."

Nikola went to go the bag of crystals. He noticed a few of the sun crystal were missing. He asked Chi Chi during the escape of his apartment from France did she drop any? Chi Chi replied "Not that I know of." Chi Chi beat up all the soldiers and they didn't have anything on them. So one of Tony's kids must have picked it up," said Chi Chi. "We will go visit tony soon, but in the meantime let's be happy." said Nikola "It will work out fine. We will get the sun crystal from Tony and when

Lord Uran attack's the earth I will defeat him." Alyssa said "After I have our baby I want to train to fight along with you and I will not take no for an answer." "Ok anything for you," said Nikola. The Justice family was grateful to get the news about Lord Uran, but Nikola looked into the dark sky and said to himself he will see the great doctor again someday.

The human spirit or some may call it soul lives on forever. Humanity is reality just borrowing a body, but will return in another lifetime or dimension. Alyssa said "We should get some sleep tomorrow is a busy day." Nikola had some teachings to give and Alyssa had a guitar lessons on her to do list. Nikola was well known around the globe. He was the man to actually stop poverty and changed the 99.9% of the people to be abundant. Nikola loved humanity he had no dislikes for anyone. This was a big meeting he had to go to. He was teaching but also had to talk to Presidents of different businesses. With every wealthy no one worked, they needed Nikola to keep their door's open also. Nikola was already ahead of the curve. He has been working on different types of robots. The beauty of it was the robot's was not the bulky looking robots. Nikola designed cyborgs and had AI (artificial intelligence).

They looked just like humans and could walk and talk, just about do anything a human can do. This made the businesses excited and they donated money to Nikola for doing this. Humanity had more entrepreneurs now so no one wanted to work for someone else. Religious believers also asked Nikola why he encouraged not believing in religion. Nikola replied "I was endeavoring to expand ignorance into knowingness." Nikola asked the lady that if God created everything, what did God create it out of? Itself? Himself? Then how could that which created us from itself ever judge us when it would be effectively judging itself. The room went into complete silence after Nikola asked that question. Nikola was not trying to persuade anyone he was just bringing it to the masses attention.

All humans were gods with the lower case letter g, but the all mighty Universe or God would be with the capital letter G. We should all love science, it is closer to the true meaning of God than any religion it ever was. Back at the Justice mansion, Alyssa was with her music instructor. She was playing a beautiful melody. The instructor name was Kaylee and she asked Alyssa what she was having. Alyssa told her she is having a son and she is excited. Kaylee complimented on Alyssa's shirt. Alyssa had on one of her clothing brand t shirts. Alyssa played the guitar really well. She played better when she imagined playing to Nikola. Kaylee was very impressed that Alyssa didn't need that much practice. It was though she was a natural guitar player.

Eight weeks passed and the family was on the way to the Hospital to deliver their baby. Alyssa was squeezing Nikola arm extremely tight because she was in pain. Chi Chi was yelling "BREATHE WOMAN, BREATHE!" Nikola told Chi Chi it's no time to joke around. Chi Chi tried to lighten up the situation. The hospital was not that far from the Justice mansion so they arrived quickly. Alyssa was checked in and took her to the back while Nikola and Chi Chi had to dress out in scrubs. Chi Chi didn't have her size but she still was able to go the delivery room. Everyone is in the delivery room, Nikola kind of nervous and Chi Chi was all eyes of excitement. Alyssa was screaming even louder pushing the baby out this time.

Nikola was holding her hand, he kept telling her he loved her and we are going to be a happy family. The baby's head came out and the doctors moved quickly after the head came out. The doctors said "Well Mr. and Mrs. Justice it's a girl." Nikola and Alyssa faced dropped but they were still SO EXCITED. They need a name for a girl. Chi Chi said she was already prepared for the girl name. The new born baby name would be Keri lee Justice. Nikola made a few jokes about three women in the house now he was go be in trouble. The hospital had a kid's store 24 hours which sold clothes. Nikola went to purchase some new clothes for his daughter. "She is the most gorgeous baby I ever seen," Alyssa said. Nikola kissed both the girls that meant the world to him. He looked worried but he tried to keep a happy look on his face. Alyssa knows her man, she said the first moment she looked at your face she knew she had to fight, just like the moment he seen the look on her face he know she was going to fight along side with him.

Chapter Nine:

After her telling Nikola that he laid in the hospital bed with his family. He told Alyssa nothing was going to come between them. The two would do anything for one another. Keri lee started smiling with so much love in the room. Chi Chi hopped in the bed also. The nurse took pictures of the Justice family. Nikola didn't let the doctors give Keri lee shots because most of those vaccine shots are pointless. Not to mention some of the shots injects babies with the chip. The chip would be planted inside a human and you can be tracked. Nikola knew that schools wouldn't accept her without vaccine shots but Nikola didn't care he was a genius he would teach her at home. Keri lee actually had two different color eyes. She had both her parents' eye color. One eye was green and the other was a hazel brown with a grey like tint to it. She was 5 pounds 6 ounces and could open her eyes as soon as she came out. Her hair was jet black and curly but Nikola notice it had a strip of pink through it. He was very intrigued that his daughter was so unique, Nikola decided to leave the hospital the next morning.

Alyssa needed to heal but she wanted to be with her family so she forced herself to get up and leave the hospital. Keri lee was here in the world now and everyone had work to do. Alyssa had a friend that was a journalist who loved a good story. She had resources that worked for the news stations. The word had to be spread throughout the country about Lord Uran attack in thirteen years. Nikola needed to train and also locate the sun crystal in France. Nikola trained for hours in his lab under his home. He was a scientist but he did every work out he had to do. His brain was healthy but he needed his body in shape also.

Alyssa was upstairs with the Keri lee. Keri lee had beautiful curly black hair just like her mother but it was still a mystery how she ended up with that strip of pink. Alyssa told Chi Chi that she thinks Keri lee likes her. Keri lee was a baby but she laughed and was not frightened of Chi Chi. All Nikola had in his mind was to protect his family. Lord Uran was a god but he still was in human form. Nikola knew he would have to face Lord Uran but the planet Jupiter usually protects the earth from comets, asteroids etc. Jupiter gravitational pull is so strong it pulls whatever is headed to the earth back and sends it the opposite direction. Uranus was more frozen ice and rock so it was heavier than Jupiter's just mainly gaseous surface.

Nikola remembered what his father taught him. His father name was OSHO. He was well known on their home planet. He use to tell Nikola mind and body as one to channel all the powers of the body. Nikola was deep in meditation for hours and hours. The home phone started to ring and Alyssa picked it up. It was another scientist from one of Nikola research centers. He asked to speak with Nikola so Chi Chi went to get Nikola. The scientist wanted Nikola to come down to the hospital because a ten year old girl had a brain tumor. Nikola immediately got dressed, and kissed Alyssa and Keri lee then left. He arrived at the hospital to see what kind of condition the little girl was in. Nikola asked the parents of the child what is her diet and does she use her cellular device a lot.

The parents said she didn't have a good diet she mainly ate fast food burgers and she talks on her cell phone six to seven hours a day. They started to run test on the girl since Nikola was the master of electromagnetic frequencies he would know more. The brain tumor was in the left side of her brain. Removing the tumor could cause death for the little girl, and Nikola didn't want to risk that. The radio waves from the cell phone most definitely caused this girl to have a tumor. Nikola was not going to let this girl suffer or die. Meditation was the answer and he knew what to do to resolve the masses who suffer from cancer. The radio waves from cellular phones were also causing cancer not to mention the pink slime in the fake beef at fast foods. Nikola started his research immediately. He needed to train and prepare himself for the earth's attack but this little girl reminded him of his daughter ten years from now. Nikola did what many men have tried to do and that was to stop world poverty.

He became even more wealthy with his technology of cyborgs robots operating. He didn't care about the wealth but he knew life should be abundant and no one should suffer. Nikola told the parents of the little girl to make her eat a better died. He gave the parents seeds from a special garden from his planet to heal every damage cell in the human body. The seeds would need to be taken with fresh pure water. He told the parents the human body would heal itself with no stress and with unlimited amounts of joy and happiness but he was not going to wait and let a little kid suffer. The parents cried ad thanked Nikola. He left out to get started training, he had much work to do. Alyssa had checked her email she had to take a business trip in Spain for her clothing line.

She had stores all over the globe but was opening a new one in Spain and she was needed. Nikola would have full time with his daughter. Chi Chi decided to go with Alyssa on the business trip to Spain for a few months. The Next day Alyssa and Chi Chi flew out to Spain and Nikola and Keri lee waved their hands goodbye. Nikola was a new dad but he did his best with his daughter. He brought her inside his lab. Keri lee was a quiet baby she never cried at all. Nikola would feed her in one arm and research and make out his plan in the other arm. When Keri lee was sleep he would meditate. Meditation and daydreaming makes a person smarter so he came up with another idea but more important things mattered first at hand. Nikola designed a copper fiber wireless ear piece for the cellular device. What this did was reducing the radio waves into the human brain which caused cancer. He also designed copper phone cases also to reduce the radio waves for the people who didn't like the earpiece.

This was a genius idea and would save lives over time. His idea was to tell people to now stay on the phone longer than four hours a day also people who owned their own businesses that require a cellular usage would surely need the copper earpiece. Nikola called some of his resources to also get a kids doll that was baled headed so little girls with cancer would feel beautiful also. These actions to place immediately with a few phones Nikola made. Nikola kissed Keri lee, he loved his daughter so much. After all his hard work and phone calls he decided to take Keri lee to the Japanese shadow puppet show. It was a movie theater in California that didn't show movies on the screen but instead it was a real live shadow puppet theater. People were under the stage with sticks controlling the puppet's which were made from wood. It was a thin screen which a light reflected off of it to actually see the shadow's of the puppets.

Keri lee loved it, to be a few months old her attention was great. Alyssa called to check on things. Nikola answered his phone but it was not a regular cell phone. He had designed a model personally for his family usage. It projected the image right above the phone and the audio could still be transmitted with or without the earpiece. Nikola talked on the earpiece because he didn't want to talk loud on top of the noise already in the theater. The image could also see the surroundings so Alyssa could see what was going on in the movies, his technology was pure genius. Alyssa had opened stores in Spain now and Chi Chi was a little home sick. Nikola said he could hear a guitar playing in the background. Alyssa was at a restaurant and on stage they played music. Nikola told her to hurry home he missed her more than anything. The two kissed each other through the projections and disconnected. The people were friendly in the theater. Nikola knew that teaching the law of attraction to the masses people would be more joyful. People just wanted to fit in and be accepted but Nikola taught a life after wealth. Because money is just energy just like everything, it's a current just like water. He wanted life to mean more than just money.

Chapter Ten:

The movie was over and leaving he noticed most people there had his clothing brand on. Nikola came up with shirts with meaningful quotes from past avatars and scientist from the 18 and 1900s, they also had different planets and constellations on t shirts. Everyone has a sun or moon sign. Nikola was a Gemini sun sign but his moon sign was Leo. His ruling planet was mercury. The universe is always working whether people knowing it or not. The next morning Nikola gets a phone call. It was the CEO of the largest toy store in the world. He wanted to thank Nikola personally for thinking of a bald headed doll so kids with cancer would feel beautiful also. Not to also mention that the cartoon developers wanted some of the action. A new character was developed with a bald head also.

Nikola was honored to have completed another goal of his. He said he is not finished yet. Nikola was currently working on a device that would detect any chemical that shouldn't be in foods. It would also reveal the percentage of beef in meat. Nikola had this technology already he just wanted it released to the public now. Keri lee started to wake up. But she didn't cry she just always looked around. Nikola had a picture of Keri lee in his lab and of Alyssa, his two favorite girls in the universe. Keri lee had gotten cranky and was ready to eat, she started to cry. Nikola left the lab and went upstairs into his mansion. Nikola was in the kitchen then he heard the doorbell ring. He goes to answer the door with Keri lee in hand. It was Jeremy he was a scientist also. He told Nikola he needed to vent and get thing's off his chest. Jeremy was having relationship problems with his girlfriend.

Nikola was not a relationship counselor but he would always be there for anyone to help. He wanted a relationship like Alyssa and Nikola but his girlfriend was really negative. Nikola laughed and said "Remember in that seminar when I said sometimes people who are together that have great age differences between them, they are together not because one is sexually attractive and one has other assets, but they are together because their souls long to be together." Jeremy said he went to sleep when Nikola was speaking on that. Nikola said "What you should do Jeremy, is write down all the positive assets about your girlfriend and leave the rest to the laws of the universe. If your girlfriend is in a mood that does not agree with the way you wrote them down about her, the universe will make you both zigzag. Your frequencies will not match up." Jeremy noticed that while Nikola was explaining that to him he had something on his mind. He asked Nikola "You miss Alyssa don't you?" Nikola smiled and said "Yes." It has been a little over a month since she left and Nikola was missing his wife. "Jeremy you and your girlfriend will do fine just think positive and have faith. The universe will attract anything you desire." Jeremy didn't stay long, he was on the way to buy a black cat because that's what his girlfriend wanted. After Jeremy left Nikola took Keri lee and they sat outback on the sun deck area. The weather was gorgeous in California.

Alyssa was calling to check on things at the house and to see how the new father is doing. She told Nikola she was staying a few more weeks. Nikola was not too happy hearing that, he was starting to sound sad. Alyssa and Chi Chi came from inside the house to surprise Nikola. She got the store in Spain up and running and they didn't need her there for the rest of the work. Nikola got up fast and hugged and told her that he missed her. Alyssa replied "I can tell." The love was once again back in the home and they could finally start building their perfect life raising Keri lee. Eight years later it was Keri lee's birthday party and Chi Chi had ordered a giant cake along with a giant princess castle. Keri lee never was around other kids much because schools educational system wouldn't allow kids with no history of vaccine shots into the school. Nikola knew those diseases did not even exist and he knew people over seas were known different from people in America. He felt it was no need for passports because once upon a time all the lands were connected and not divided.

Nikola taught her at home, he knew more than most teachers. Everyone started to sing happy birthday to Keri lee and she smiled. She stood out over the other kids, her dark hair shinned in the sunlight and she had more pink in her hair now than when she was born not to mention her eyes were two different colors. Nikola kissed Alyssa and said "I love you so much and our daughter has gotten so big, she is beautiful just like her mother." Alyssa had gotten Keri lee a guitar for her birthday, she started to play it soon as she opened it. Nikola didn't even know his daughter could play the way she did. She could play anything kind of music a guitar could play but it was her first time playing. The kids loved Keri lee even though they thought she was different because of her eyes two different colors but Keri lee was very confident and highly intelligent for her age. She didn't worry about what others thought. She was brought up in a happy family with love and abundance so that play a big part of her life.

Chapter Eleven:

The party went on for three hours and Keri lee actually wanted to get back to her studies. Every since her birth she was always in the lab with her father learning while observing. She loved science and astrology just as much as her dad did and could understand everything he worked on. It was time for Alyssa and Chi Chi to work out and the Justice mansion has an exercise room so it was no need for traveling to a gym. Keri lee and Nikola went to his lab to start teaching Keri lee todays lesson which consisted of the equation E=MC squared. Keri lee said "That means everything is energy." Nikola was shocked she knew that and was proud of his daughter for actually paying attention. He explained everything to her that school systems do not teach. Nikola had an image on the wall of a human brain and Keri lee named each part of the brain. She told Nikola that the corpus callosum is what bridges the left and right brain together. She seen where Nikola had on his chart that humans only access ten percent of their brains and the other ninety percent is unknown.

Keri lee asked lots of questions. She asked her daddy what if a human access one hundred percent of their brains? Nikola said "It would be a miracle and humanity is not ready to access that right now. They may be looked at different and people would be scared." He continued and told her "That's why we pursue knowledge without any interference of our addictions because doing that we would not manifest knowledge in reality and human bodies would experience new chemistries, new holograms and new elsewhere of thoughts. It is a term people say, that you can do anything once you put your mind to it. If the human brain can use ninety percent the humans could fly and honestly do anything they desire just on a thought." Keri lee knew that her dad was a god and he genetically engineered humans. She asked "Dad, why not genetically engineer humans to use one hundred percent from first created?" He told her "We could not access the neurons witch controls the human electrical thoughts. The human brain will change along with the earth. Meditation and yoga tunes the human brain but we will discover in time ways to do it faster."

The lesson for today was over and Keri lee went to her room to play her guitar, she impressed her dad already with her genius. Keri lee would also practice fighting in her room with deep meditation. Alyssa had taught her different types of fighting styles but she never told her dad she knew how to fight. Of course a father is not going to want his daughter to fight but Alyssa strongly loved her family and she wanted everyone on the same levels so she helped her whenever she could. Four years later Keri lee was twelve years of age. Nikola said "It's time, we have one year until Lord Uran attacks." Nikola made several attempts to call the hotel his friend Tony in France was staying at. He never got an answer just the voicemail. It didn't matter too much to Nikola, he surely wouldn't want to go get the crystal in hand anyway plus Nikola wouldn't trust getting it mailed to him. Nikola told Alyssa to contact all her resources and her friend Naomi at the news center to spread the word over California. The media would just have to tell the masses that it will be a comet so that people wouldn't panic, but the comet would be crushed by the earth's gravitational force. The comet would be headed towards California and there is nothing to panic over.

Nikola kissed Keri lee and Alyssa before he and Chi Chi would leave for France. Nikola told Keri lee she has gotten so big now from when she was just in his arms twelve years ago. Nikola and Chi Chi set out in his spaceship to France. Keri lee asked her mother could she go stay at Jenny's house one of her friends from the birthday party. Alyssa told her "Yes baby go get your bags." Alyssa had work to do and she wanted Keri lee to have something to do. Alyssa pulled up the super computer to do multiples things. The windows all over the house were screens for the super computer also Alyssa could project the image right before her eyes. Nikola had all the technology but didn't want it to get out into the wrong hands. Jenny's mom had arrived to pick up Keri lee and Alyssa walked her outside to the car. She took her guitar because she loved music. Alyssa played the message Dr Borosky left them but like always the news would relay false advertisement. The media always say if it is good news it's not news. Not to mention they will leave off ninety percent the truth. They said it was a comet headed towards earth and its effects could cause a power outage.

The military was contacted to have them on stand by just incase things got out of hand. She called Nikola and told him the news that she got out. Nikola was excited and told her they were halfway to France. Nikola said he loves her and he will make it back safe into her arms. Keri lee was having fun at jenny's slumber party. A cartoon from 1980s came on television and that caught the girls' attention then the girls were pretending to be rock stars. It was a cartoon called KEM and it was about a group of four ladies who played different music instrument and traveled the world. Keri lee started to play her guitar and the girls loved it. Jenny's mother came in the room to hear Keri lee play. Jenny asked her mom could they color their hair like the girls on television. The mother answered yes and she left out to the store to purchase the hair dye. She called Alyssa first to ask could Keri lee color her also it was washable so it would come out in water.

On the way to the store everyone stopped and looked at the news. Keri lee had seen her mother on television and the news was saying that Dr Nikola Justice wife Alyssa Justice had predicted a comet to come in the earths orbit this year. Keri lee said "That's my mother on television and that couldn't be right. My dad wouldn't just leave if something was coming." Keri lee was already educated on the media just telling under fifty percent the truth. Nikola finally made it to France and headed straight to the hotel where Tony said he would be staying at. They arrived to the hotel and walked in the entrance, Mohini was working the front desk. She was so happy to see Nikola again. She said "The kids are fine just gotten bigger now but Tony went out to the marketplace." Nikola told her he tried to call the hotel. Mohini told him it is hard to get someone to answer calls when she is not working the front desk. They needed direction to the supermarket to asked Tony about the crystals. She gave them directions and Nikola told Chi Chi he surely hopes one of Tony's kids have it. The marketplace was not that far from the hotel so Nikola and Chi Chi went quickly inside. Tony was happy to see them. Nikola said "Your wife told us where you were at.

He asked Tony has his kids had any crystals? Tony called to the kids then Nikola seen the crystal around Jin's neck, he was wearing it as a necklace. Chi Chi looked at Nikola they both knew that the boy had bonded with the crystal. The ancient Sumerians collected crystals from the gold mines in east Africa when they first came to the planet earth and the crystals were used to remove karma, to heal or to obtain more wealth. Each different crystal had the powers from each planet or star in the universe. The reason the crystals was cast to the planet earth is because planet earth is mainly water and water hold a profound amount of energy with thought also. Whoever bonds with the crystal will become a part of it meaning you can not be separated from it so that means that Jin was the master of that crystal and the crystal would need to be placed on top of the great pyramids in Egypt.

Chapter Twelve:

The senator in France contacted Lord Uran and told him that the governments were expecting an attack and also Nikola had accomplished his goal. Nikola taught the world the laws of the universe and astrology. The people didn't live 90% negative anymore and the world has changed completely. Lord Uran was furious he couldn't wait the full thirteen years to head to earth. Lord Uran came from beneath the depths of the frozen ice on planet Uranus and he summoned his fleet from Uranus twin planet Neptune. Both were bluish greenish planets with similar elements. He didn't want humanity to think outside the box because with out fear there was no control. Lord Uran departed and would arrive to earth in a few hours. With him entering the earth the earth's north and south poles would reverse, this would cause weather disasters and earthquakes. Chi Chi asked Nikola didn't he feel the earth shake. The earth was vibrating and the clouds were standing still. Nikola could tell the direction the wind was blowing and it all of a sudden shifted the opposite direction.

Nikola told Chi Chi Lord Uran must have decided to come early. Nikola explained to Tony what the power of the crystal does. Jin couldn't even remove it which was awkward because the sun crystal couldn't be touched by no one else only by the owner of it. Tony was asking what was going on and he didn't want his son to have to leave to Egypt. Nikola said "It has to be done or we won't have a planet to live on." The spaceship would only seat two adults, so Tony couldn't join them. Jin understood his destiny because the crystal showed him visions and Nikola answered a lot of his questions. Jin told his father that he will return soon and to not worry he is older now. The three took off and set direction to Egypt. Nikola called

Alyssa to update her on the news and to warn her that Lord Uran was on the way. He asked her where his daughter was, Alyssa told him over her friends house. Alyssa called for Keri lee to come home but no answer from Jenny's mom. Alyssa told Nikola she was going to keep calling to see where Keri lee is and told Nikola to make it home safe. Chi Chi asked Nikola what was his plan fighting Lord Uran? Nikola said he has not thought of a plan yet. She replied "Win or lose I'm with you."

The closer they got to Egypt the faster Jin's heart beat was. Nikola told him it must be the power of the pyramid reacting to the crystal. The crystal started glowing and Jin's pupils turned white. Chi Chi told Nikola to land because Jin was in pain. Jin said "No don't land let me out," "Let you out!" Nikola said, then Jin opened the doors and jumped out. They were not at the pyramids yet but Jin could some how fly now, the power from the crystal had to be channeling through his body. They followed Jin but he was super fast and Nikola was flying the ship fast as it could go. They couldn't loose Jin because he was all that stood in keeping the earth together. The earth all of a sudden starts vibrating multiple earthquakes, tornadoes, and hurricanes started. The entire state of California power went out and Alyssa was trying to call and find Keri lee but even the generators Nikola had under the house was not working. She tried to leave and head to Jenny's house but the cars were not starting.

Lord Uran has arrived onto the earth plane and his presence onto the earth is the cause to all these disasters. Nikola was worried because he had to make sure the crystals were in place to get power from the sun. Worldwide it was though the earth was crying in pain. Jin made it to the pyramid, Nikola told him to align himself on top of the pyramid. Nikola started doing the sun moon technique that was the secret of the crystal to access the full potential of the crystals. Nikola had to visualize the sun and the moon merging together into his mind and heart. Jin started glowing and the pyramid shot a beam of energy into space. The entire universe is connected so the sun transmitted energy towards the earth. Nikola said "Its working, the sun's energy was powering the earth Jin is the power source." He couldn't move from the pyramid but at the same time Lord Uran arrived above Nikola's mansion. He started freezing his house with no one inside but Alyssa.

She runs towards the lab downstairs because she remembered Nikola kept an emergency phone that ran off solar power but could only connect with another solar power source. She calls Nikola phone in his spaceship screaming because Lord Uran was above their house. Nikola said "Chi Chi we have to go." The speed of the earthquakes had decreased because the sun was charging the poles of the earth now but it was still not good for the earth the longer Lord Uran was on it. The military started opening fire on Lord Uran spacecrafts. The military had solar powered tanks and jets also. A loud frequency burst came from Lord Uran ship, crushing the jets and tanks in seconds. At this time Nikola didn't care about using light speed to get to his wife he just needed to get there. The earth would start to shake again with him doing that because this speed would come from the universe where time does not exist but on a planet such as the earth it would do massive damage.

Nikola had to do it, because it would take him over six hours to get to America from Egypt and using light speed it would take him under twenty minutes. After Lord Uran destroyed the military tanks and ships he landed and approached Nikola's home. Alyssa came out and asked Lord Uran what did he want? As he floated over head he asked "Where is Nikola?" "My husband has no business with you." Alyssa was ready to fight for her husband. Lord Uran asked Alyssa "Did you not observe how easily I took out the government? Your precious US government, the strongest government in the whole planet crushed in mere seconds yet you dare to stand up to me, a puny human. Ha, We tell earthling jokes where I rule but I see your husband cleaned up all the corruption that I started." Alyssa asked "Why do you want to enslave the human race?" Lord Uran answered in one word......POWER.

Keri lee was in the car with Jenny and her mother and the cars still was not able to start. They took off walking to safety, but Keri lee could see ships floating above where she lived. Jenny's mother told her she couldn't walk in that direction to go inside the store with the others until this blew over. Keri lee said "No that's my mother and father at home." Jenny's mother said your "Going to get yourself killed." Lord Uran asked where Nikola was again but Alyssa was silent. He approached her and grabbed her neck squeezing it. She was resisting him but he over powered her and he threw Alyssa into the door. A large ball of electricity formed in the sky, it was Nikola and Chi Chi. the ships started to open fire upon Nikola ship but he started firing back headed directly towards Lord Uran. He crashed his spaceship directly into him. He damaged his ship and Lord Uran was pushed back. Nikola headed straight to Alyssa but she was unconscious. He put her in his arms, she was still alive.

Chapter Thirteen:

Nikola was now upset and ready to go up against Lord Uran. The earthquakes started back and tornadoes were even stronger because Nikola used light speed. California was surrounded by water and the ocean waves were coming onto the beaches. Lord Uran told Nikola "You ruined all my plans. You actually brought peace to humanity. We kept the secret to life hidden (law off attraction) for decades." Nikola said "It should be known to all the entities of the universe not just a few." Nikola told Chi Chi to stay with Alyssa then Nikola and Lord Uran started to battle.

Nikola punched him in the face and chest but he was not hurting Lord Uran. Lord Uran tells him "Earth and love has made you weak." "My father sent you to Mars to die and I will send you back," Nikola told him. This made Lord Uran mad and he kicked Nikola directly in his head. This knocked him a few feet back. They were punching each other so hard all you seen was electricity on their fist. Nikola was fighting out of love which was stronger than hate which is what Lord Uran was fighting for.

Jin would need to concentrate more energy to keep the earth together. Outside of Pluto was another solar system with a sun a million times larger than the earth's sun. The entire universe was energy so it can be transmitted. He used energy from that sun to help keeping the earth together. Alyssa had got up and seen the fight. She wanted to do something to help, because Lord Uran started to get the upper hand. Nikola let his guard down paying attention to Alyssa an Lord Uran punched and it knocked him out. Chi Chi and Alyssa ran to his aid, Chi Chi flew up behind him and kicked Lord Uran in his head which gave Nikola the chance to hit him with a blow to the face but Lord Uran turned froze Chi Chi solid and hit Nikola with a head butt. Alyssa was all alone and she was fighting, even though she was no match Lord Uran toyed with her. Nikola got up slowly, he was hurt. His arm was broken and his body badly beaten, Nikola was in bad shape now. He ran to save his wife and quickly Lord Uran dodged Nikola and froze Alyssa. His eyes opened wide he couldn't believe this but Nikola had heart he was not giving up. He found with one arm and still had his legs and mind.

Tornadoes started to head towards the battle and things started looking bad for Nikola. Lord Uran punched him towards Alyssa so they could die together. He seen the tornadoes coming so Lord Uran said "I didn't have to kill you after all, I will just freeze you and let the wind crush you both." Alyssa hand was outside the frozen ice, Nikola held her hand and told her he loves her. The couple was helpless on the ground. The ships surrounded them in the air and Lord Uran was just about to freeze Nikola, until he heard a soft voice yell "FATHER." Keri lee had made it home to see her family in pain on the ground. Nikola told Keri lee to run away. Lord Uran seen Keri lee, he turned to Nikola and asked "You had a child?" Nikola yelled "LEAVE HER ALONE!" Keri lee walked towards her family.

The tornadoes were coming directly towards them. Nikola was crying and told her to get out of here. Keri lee seen her mother trapped in the ice. She walked over to her mother and touched the frozen ice and it started to melt. She did the same for Chi Chi. Nikola was looking puzzled because he did not know she had powers like this. Lord Uran ordered his ships to open fire Keri lee controlled the tornadoes and sent the wind towards the ships. The wind left no traces of the ships, everyone was looking confused. Keri lee floated in the air towards Lord Uran, and Lord Uran sensed her massive energy.

He said "It can not be." Keri lee could see nothing but energy wherever she loved. Everything was energy. She removed the ice powers Lord Uran had and he fell towards the ground. She had to move quickly because the earth was still falling apart due to his presence. She used his powers against his and turned most of his body into ice leaving only his head exposed, she then told her parents goodbye and she loved them. Nikola said to her "No Keri lee, what are you about to do." she grabbed Lord Uran and flew up towards space. Lord Uran had to be destroyed so she through him directly into the sun. Keri lee had the powers of Saturn and she didn't need air to breath in space. Since everything was energy her Nebula light (soul) energized her lungs when air was needed.

Jin became unconscious once Lord Uran inside the pyramid. The pyramid pulled him inside. Keri lee could light speed through space with ease, she was intangible. Keri lee made it to the sun with Lord Uran to destroy him. His power was too great to be dealt with on the earth. Keri lee knew how energy worked and that it never can be destroyed only changing but it wouldn't be inside the earth atmosphere. The two exchanged words as she flew him to his demise, she threw him directly into the sun in seconds it was not a trace of him left. Keri lee remembered what her father said that she would be looked at differently if a human could access their full potential of their brain. She didn't go back to earth, she wanted to explore the universe and experience her powers.

Keri lee always thought that the universe (heaven) was always black but she actually noticed it was beige like color. Alyssa asked Nikola "Do you think she will ever come back?" Nikola replied "She has to, she is our daughter and we love her." Nikola looked up into the sky and mentally said "Thank you Keri lee, we love you dearly, please come back home someday." His spaceship was damaged so he couldn't go see how Jin was doing at the pyramid. Back in France the senator knew about Lord Uran defeat he was leaving France, but outside the building was Tony with a gun pointed directly at him. Tony shot the senator in his head saying "This is for my son." Tony was not sure if his son was dead or alive but he knew that the Senator was pure evil.

A week later, Alyssa wakes up in the middle of the night breathing hard and fast. She has been having the same dream every night since the battle with Lord Uran. Nikola said "baby what is it?" she answered "It's the same dream over and over, of me and Keri lee running towards one another and when I get to her I go right through her and fall. It is like she is a ghost and we go through one another. I go through her, fall and just lay on the ground begging her not to go." Alyssa got up and went to the balcony outside their room and looked up at the stars. She asked Nikola what does he think Keri lee is doing up in heaven. He said he did not know but she is a kid, her curiosity becomes her. She will come home honey I'm sure of it.

Chapter Fourteen:

Nikola and Alyssa spent most of their time in his laboratory repairing his spaceship. Alyssa kept having these strange dreams ever since Keri lee has been gone, and for some strange reason she felt something was telling her to go in space. Alyssa helped with the repairs because Nikola arm was still in a sling from being broken by Lord Uran. Chi Chi wounds healed quickly and she was going to stay behind incase Keri lee comes home. They were prepared for this new journey and started to take flight. Alyssa told him to head towards Saturn. Nikola asked, "Why Saturn?" Alyssa told him "it's this feeling in my soul that's telling me to go to the moon of Mimas." The repairs to the spaceship would allow them to use light speed even faster now. They arrived on the Mimas and the first thing Alyssa did was smile because she knew Keri lee has been here. They looked around and seen footprints on the ground. The footprints could make out to be human footprints.

They continued to walk around looking for traces of Keri lee, but they came up with nothing. Alyssa head starts hurting and it was a throbbing like pain. Nikola asked her was everything ok? She replied, "Yes! They continued to walk and Alyssa started getting weaker and weaker until after a few minutes she passed out unconscious. Nikola picked her up and called her name. "Alyssa! Alyssa! Wake up!" She was unconscious, but Nikola didn't know why she had passed out. Alyssa was in a deep sleep but mentally she was aware of her environment and she could see Nikola but he couldn't see her. She was looking at him calling her name and holding her trying to wake her up. She started to cry mentally because she seen him crying and screaming her name and she wasn't able to move her body.

She was actually in another dimension. Nikola kept talking to her saying "Don't leave me, don't leave me." He never loved anyone or anything as much as he loved his family. Alyssa was physically in another galaxy but she could hear Nikola voice. She was in another solar system light years away from her home. This solar system had a larger sun, four billion times larger than the earth's sun, so it was extremely hotter on the planet she had arrived to. Saturn profound aura protected her no matter where she was in the universe so she wouldn't burn. The planets here could with stand the heat. It wasn't anyone in sight all Alyssa seen was water. She looked down to her feet and noticed she was standing on the water surface. She could tell it was hot because of the heat waves in the air but this was a new experience to her. She also notices the storms, it was raining behind her and snowing in the other direction.

The rain wasn't hitting the ground it was just dropping from the sky. She continued to walk and approached a new sand like surface but similar to grass. Alyssa laughed and had no clue where she was, but she knew she had to leave. A forest was in sight now and Alyssa headed towards that direction, and she started to hear cries. It was a group of kids no older than ten years of age. They were human like beings just like Alyssa. She asked the kids where their parents were. The six kids looked at Alyssa and started to laugh, and Alyssa was curious at what they were laughing at. The kids knew Alyssa wasn't from their planet so they surrounded her and started touching her and pulling her hair and observing her. They didn't talk but Alyssa noticed them communicating but she was confused on how. Alyssa was curious now so she started mentally thinking about things.

She tried to speak to one of the kids, not say it out her mouth but in her thoughts. She spoke and noticed that the kid she was speaking to looked at her smiled and spoke back to her in his mind. Alyssa was amazed how it was no language on this planet but she could hear the kids' voices in her head of whoever she was talking to mentally. She asked the kids where was she? The kids told her she was in the celestron galaxy; it was billions of years away from her home planet. Alyssa asked them how she got here. They told her they didn't know but their grandmother could tell her but she was in the city. Alyssa told the kids she needed to go there but the children told her no they couldn't go.

The boy she first spoke to was the oldest he introduced himself to Alyssa and asked her to have a seat. The boys' name was Bobby and he introduced the other five kids to Alyssa. "This is Keko, she is nine and she was born def. Her twin brother on the other hand was not, his name is Angelo. They have a special ability to heal each other but only if they are near one another. Next there's Aden, he is six the youngest, also the one in the wheelchair. He has the ability to see clearly even in pitch black darkness but he lost the use of his legs when he went out to help his dad hunt. These things called Argos are vicious creatures that have razor sharp spikes on their backs and eat anything. As they were hunting one came after them full speed and his dad jumped in the way saving him, Aden ran and tried to get up the tree but the Argo scratched at his legs trying to pull him down and cut the muscles in his legs. He made it up the tree but he couldn't move and the Argo circled around waiting for him. The next morning another hunter who was friends with his dad found Aden and what was left of his dad and rushed Aden to the local doctor, he lost so much blood they thought he wasn't going to make it. Over there is Lili she is seven, now we all talk threw thought as you have seen but she can move things with her mind. She just now getting real control over her ability, at first she would make the things in her room float around while she was sleeping and when she had a bad dream she would break doors and windows trying to escape. The last time she lost control of her powers she almost killed Amari her older sister. She is eight and lucky for her she can phase threw things so when Lili lost control that night she phased threw the bookshelf that was thrown."

Alyssa asked Bobby "What's your story?" "Well with me I have the ability to generate fire." Said bobby, "But what happened, did you loose control one day and burn down a tree or something," Alyssa asked. Bobby looked at her, took a deep breath and told her what happened. "Some thieves tried to rob me and my mom, they took her bag and pushed her down at the same time one was holding me back making me watch. He had the ability to control others minds and he made my mom tell me she hates me and wish I were dead. I lost it and I killed the one that was controlling her mind, the one that was holding me let go and looked at me in shock. I was about to kill him too but my mom grabbed me and said that's enough please don't kill him. When I came too she was crying and from that day forward I have kept my temper under control swearing to never kill again."

Chapter Fifteen:

Bobby told Alyssa that they couldn't go back to their homes was because they were indigo children. Indigo children were born with seventy percent of their brains activated and could do anything that comes to their mind. Bobby and the other five indigo children were the last six in the entire celestron galaxy. They were raised to be evil and rule different planets. The grandparents of the children attempted to create love and peace to the children but the ruler of this planet had other plans for the children. The king of this planet wanted the kids as slaves or wanted them extinct. "These children were much like Keri lee," Alyssa thought to herself. She thought out loud so the children asked who was Keri lee? Alyssa laughed because she forgot her thoughts could be read.

Alyssa knew these kids had to be very intelligent to make it on their own like they did because it looks like no civilization out here. One of the kids was in a wheel chair, his legs were crippled. After he read Alyssa thoughts about him being crippled he told her he could out think her, sometimes a physical impediment brings out the greater mind. Sometimes the uglier face and the more common the body the more brilliant the mind. The body becomes a distraction in evolution. The brain is inter dimensional and has a fabulous characteristic of being able to remain in any worlds simultaneously. Every entity can think in inferred and that thoughts are levitations of the future. Alyssa knew these kids were brilliant because they say the same things she heard Nikola say in his lectures. Alyssa told the kids she was taking them back to town and not to be scared of the king. She also told the kids if they are as powerful as they say then they can control their own reality.

The kids shouldn't have to run away from their loved ones, but sometimes life changes. Bobby asked Alyssa did she lose a loved one? She said her only daughter Keri lee left her planet earth and now she was in search for her. Alyssa would not stop until her daughter was found so she knew how the parents of these lost kids felt. She told the kids she is returning them home to their families. Bobby told Alyssa they have to travel through the valley of souls but the people come up missing in the mist. The kids had by passed this area because it was a sewer tunnel before the valley but its only kid's size, so Alyssa couldn't fit in it. Alyssa was very brave and fearless because her destiny was to return back to her family. Nikola tried to pick Alyssa's body up but his arm was still in a sling but he didn't give up. He needed to get Alyssa back to earth. The winds started to get high on Mimas so Nikola grabbed her arm and struggled to lift her with his one arm. The high winds came from nowhere but they were too strong for him to stay there. He summoned his ship to come to them. The ship came to his need and he lifted Alyssa into the car.

Alyssa asked "Why do they call it the valley of lost souls?" They told her it was for soulless beings that never thought for themselves they only took orders from a king or whatever high ruler that controlled them. Alyssa said "So they're something like slaves or peasants?" Bobby said "Something like that". They arrived to the valley and Alyssa seen pirate ships that were sitting on top of the fog. Then she noticed they all were walking on the fog. This planet was nothing like she ever experienced before. The kids were fearless to be their ages, but the group proceeded forward. Beams of energy surrounded Alyssa and transformed into beings. They could speak from their mouths unlike the children. Alyssa said they don't want trouble they just want to pass through the valley. The leader of the soul pirates introduced himself as Seth The Great.

He was an ancient warrior decades ago and was banned to this soulless valley because his hatred. Alyssa started to get a picture of why these soulless beings were here; they lived a negative hateful life. Alyssa asked Seth, "Why did you hate the world so bad?" Seth told her he loved a woman before and the woman of his life died, her name was Elizabeth! Seth ruled the Milky Way galaxy decades ago trying to get the entire universe to be one with love. The entire universe is a giant magnet and everything was energy. Alyssa knew that concept because Nikola master the equation E=mc squared. She asked did they know of a man by the name of Nikola Justice. Everyone had blank looks on their faces, and said "We know him. Nikola was a legend!" She said that's her husband, and she is trying to get back to him. Alyssa was just a human in this galaxy but everyone could catch her wavelengths she was inspiring and determined.

Seth told her in his ship he had a time travel machine to get her back to whatever dimension she came from. Alyssa smile and was ready to leave, but she remembered she promised the kids she would get them home safely. Seth said "The time machine is charged now and it will remain charged for two more hours but after that it takes three months on this planet to charge." Alyssa took her chances on faith and destiny, and asked Seth to let go of his past life and start over, life is not a test it's an opportunity. Alyssa told them all to start governing their thoughts and it is not too late to change their lives. We only live in the now and its no right or wrong path to life. We are only here to live the experience nothing more. Everyone was impressed to see a woman so intelligent give them speeches like that.

Alyssa definitely made a believer out of herself, and the pirates started to listen. One of the kids said that the pirates could live in their city and protect them from the king. Bobby explained to Seth about indigo children and how the king that rules over their land wants to end them. Seth was not happy for that, and Alyssa was happy she motivated the soul pirates to change the way they were living. Love is the most powerful emotion in the entire universe it is that God feeling and it conquers all. Seth went from a life of hate to a life of peace and love. Alyssa noticed a few of the other soul pirates holding hands, as they were lovers. It was actually two men holding hands and two women. The two men fell in love with each other before they were lost souls but some told them that they can not be together so they lived the rest of their lives hating the people who were in love with their so called soul mate. When the two had enough they ran off together into the valley and ran across Captain Seth who took their souls forever. The two women were friends who hated anyone with children because they could not have children of their own. Since neither on could be a mother they promised each other they will always be together even in death. One day they left to find a place where there are no children and ended up in the valley. Captain Seth found two more lost souls to keep forever.

Alyssa didn't question it, but she looked around and noticed that no one on this planet judged anyone. The kid in the wheelchair was crippled and no one picked with him and it is gay couples even as spirits and no one judged them. Everyone took off towards the city where Alyssa can get the kids home safe and stop this kings attempt to assassinate these children. The city was beautiful when they all arrived;

it was a giant water city. The houses were made of water and the trees also it was paradise. The kids were happy to be home and ran towards there houses.

The parents of the children came out hugging their children. Alyssa started to cry seeing all the love she just brought besides she missed her daughter. Bobby told his grandmother who Alyssa was and the grandmother could speak just like Seth. It was only the kids who talked mentally, through their minds. The grandmother said it's not safe for the children to have returned, Alyssa took the blame for the children returning because she knows family is important. The objective now was to visit the king and Alyssa was on the way to do so. She walked passed all the guards no one touched her, it was obvious the king was waiting for her. Alyssa made it to the kings thrown and everyone kneeled but Alyssa stood. He was king Phoenix; he took over after a great war on the planet took place. It was a battle between soldiers and indigo children. The only remains of the indigo children are the six that's in the city.

After the Indigo children turn into teenagers they are none controllable and cant be contained. They listen to no ones rule and they have the power to destroy a civilization but as a youth there powers are stable. Phoenix didn't want to hear anything Alyssa had to say, but he made her a wager that if she takes his side on the thrown he would spare her. Alyssa said she has fought gods before and she has no fear in her heart. She told phoenix about the death of Lord Uran. The king said he would spare Alyssa because she reminded him of his past queen. Alyssa told him he needs to leave the land because it's not right for him to destroy kids and also rule over people. Everyone should have their own rights and controlling people is not one of them.

Chapter Sixteen:

Seth and the pirates were right behind her so Alyssa had plenty of help. Alyssa said fighting was not always the answer but she would fight for what's right. The king didn't want to battle with Alyssa or the pirates but he mainly looked at the aura around Alyssa that's what had him nervous, it looked as though it was a face. The king actually told his soldiers to follow him and he left the castle to take over another land. That was very awkward to everyone but Alyssa thought it was her positive thinking and even maybe Seth scared him. The children ran with smiling faces towards Alyssa, the entire city started to gather because the king had left. The children could have defeated the king but Alyssa didn't want to battle unless it was forced upon. Seth told Alyssa its no way she can get home through his time machine because a few hours have passed.

Alyssa was upset, until Bobby walked up to her and mentally told her that his grandmother knows a way. Alyssa went back to the grandmothers' house and she understood Alyssa situation. The grandmother told Alyssa it was a mystery to her why she came to their world but she was happy that everything worked out so well. Alyssa needed to head back to her dimension, she said she needed a time travel machine to leave. The grandmother told her she didn't have any of the later technology, she never needed it. Everything you ever need will always be in the depths of your brain. Alyssa didn't quite understand but she had a clue what she wise old woman meant. Alyssa went to talk to the children and Seth to say her goodbyes. She told the kids she was happy to have helped them and told them when they get older to spread love and joy to wherever they are. Seth asked her is she sure she didn't want to stay? Alyssa told him she had a family that loves her and that was waiting on her when she awakes.

Alyssa needed to meditate now, so no time machine was require. Her brain could do anything she puts her mind to because the universe is connected at all times. She sat in a yoga posture and focused her mind on one thing, her family and returning to them. She could start to feel Nikola and his love for her. She smiled in her meditation but she couldn't feel Keri lee energy. The aura around her body started to glow brighter and she started to levitate. Everyone waved her goodbye, she went though time but she started seeing her dreams again. The same dream over and over, but she knew Keri lee was out there.

She returned back to her unconscious body and awakened. Nikola was passed out in his spaceship. Alyssa knew how to pilot the ship so she started to head back to earth. The steering wheel was wireless so she could sit in the passenger seat and fly home. Nikola woke up and jumped when he seen Alyssa, he asked her what happened? She said she had the adventure of her life. They kissed one another and were happy to be side by side again. They headed back towards earth, the couple was exhausted. Alyssa told Nikola she has a new idea and it will be quite prosperous. Four months later in Miami Florida, Nikola and Alyssa were visiting the new club they had built. Alyssa had come up with an idea while spent her time in the celestron galaxy to open a club. It was no ordinary club but a club with a restaurant and six clubs and a comedy stage build in a 20 million dollar building. One of the clubs had an indoor beach just incase it rains in Miami or if it gets to hot it's an indoor beach.

It was a gay club linked in with the other clubs and this was Alyssa idea to link the clubs together because she noticed the gay pirates not being judge so she wanted humanity to all get along. Every entity was under the same blue sky as everyone else so everyone should get along, no matter what lifestyle someone lives. Marijuana was legalized in Florida so it was ok to smoke in the club. Nikola didn't smoke but he always states that you can't cure an addict until you give an addict everything they want and then ask for no more that is when the individual has owned an experience and become wise. The club would generate six million dollars a week, this was a brainstorming idea. The only club in Miami open 24 hours, they named the club No gravity. Nikola named the club and basically everything he does has something to do with science or his family.

Chapter Seventeen:

Nikola had real people who actually were more than happy to work at his club, no cyborgs were needed. The beautiful parts about the cyborgs were that they had human intelligence, so it was though it was a real human. Back in Egypt Jin was living with a family there who seen him lying the sand months ago. Jin slept with no energy for months but he finally awakened and was able to walk now. He looked at his neck and the crystal was gone but actually the crystal was a part of him, and it was an imprint on his neck. He also looked at the markings the crystal put on his body. He thought that the power would be too much for him but the ancient markings on his body told a different story. Jin knew his family was worried about him, and he needed to return to France. The people of the pyramid were really friendly to Jin and took him in as their own, but he needed to head home.

Jin noticed he was inside the pyramid still because it was another pyramid inside the pyramid. Jin said his goodbyes he had a long way to return home. After Lord Uran's death the elites over the globe had no choice but to back down. One man can make a difference but a million men can change the world. Nikola made a way for humanity with his teachings but the people also played their role. Nikola started programs for people who had mental issues. No matter how much money someone has if their not happy they will never be free. Nikola explained that money doesn't always bring happiness but it does bring freedom.

After humanity stood up the government was still there but was powerless. Nikola didn't want to control anyone or power he just always wanted people to acknowledge the great intelligence inside themselves. Nikola didn't put money in everyone's pockets; the power always was there he just brought it to their attention. It was counseling for rapist and murders because they may have had money but Nikola believed in each individual. Nikola opened more schools up teaching the youth about more science and physics.

He funded everything himself schools, space camps, and counseling. People had relationships issues also but Alyssa told the men and women to write down all the positive aspects about their lover and trust that the universe will bring you whatever you visual. Nikola created his entire reality from just a thought that took place on the moon of Mimas years ago. He attracted his home, Alyssa, and all the luxuries he can endeavor now. Hospitals lost more patience because everyone was living an abundant life so the body was doing what it supposes to do, it could heal itself. Nikola knew stress does damage to the body and years ago most of the people was stressing and focused on how they would escape this debt.

Women and men who have a predisposition of blindness could turn it into a predisposition of sight, it was beautiful and surely heaven on earth. With this much positive energy and love on the earth plane the water streams started to manifest a electrical charge in it and it actually destroyed all the fluoride in the water which was poisoning people. The earth is a living organism the same as human beings, so it was beautiful to see the trees a brighter green, the skies a brighter blue. The earth started producing faster and juicer fruits and vegetables. The vegetables were so good now that the social habit of salt had decreased. Salt throws the human body into a imbalance, it can't decrease the toxins through the urine if the body has to much salt in it, water can't get out to cleanse the body.

Humanity had finally unlocked the mystery of the ninety-nine percent junk DNA that doctors could not figure out. People didn't focus on their emotional means to an end anymore, and the recycled ignorance would stop. Nikola schools taught every science and it was more hands on and not just lectures, and had many different levels also. The lower levels were for the youth and the upper levels were for relationship counseling, and child services for kids with no families. There were mother and father programs that paid child support and didn't get to see their children. This wasn't right at all, and Nikola was big on family. With Keri lee gone Nikola was really serious about families staying together and would help any man, woman or child who needed it.

The courts around the nation favored what the people wanted because the people hold power and governments had no choice but to. So it was ordered that if you paid child support it was by law now that you will see your child no matter what the other parent had to say. Alyssa was great on the guitar, so she told Nikola she wanted to start a band. Nikola laughed and asked was she serious? She looked at him with a mean look, and told him yes she was serious. The current guitar she had was a Gibson Es0335 TDC, but she wanted something more stylist and also a louder sound. She started searching over the web for guitars she may have in mind. What she had planed was to get a general idea and get one custom made.

She found a Stratocaster guitar played by Jimi Hendrix that gave her an idea of what she wanted. She needed specs that consisted of a body: left handed two – piece select Alder body with reverse "belly cut" contour on front of body. The neck of the guitar would be one piece C shaped maple neck with reverse, large, 60s style headstock (vintage tinted nitrocellulose lacquer finish) with a fret board: maple, 7.25 radius (184mm). Scale length: 25.5 (648mm), Frets: 21 Vintage Style Frets , Width at Nut: 1.659 (42mm), Machine Heads: 80s Style F Stamped Neck Plate with Black Plastic Gasket, Hardware: Chrome, Pickups: 3 Custom 69 Single Coil Strat Pickups with grey ribbon, Pickup Switch: 3 way, Controls: Master Volume, Tone (neck), tone (mind), Pick guard : 3 ply white/black/white, Bridge: Vintage 70s style cast tremolo bridge with cast saddles.

She contacted a friend of hers that owned the Guitar Center and gave the specs and told her she wanted a very bright green color. Alyssa now had to start searching for talent and she got Chi Chi to help her do that. She needed a drummer, bass and someone on the keyboard. Alyssa was going to be the singer and lead guitarist. Chi Chi called the news stations and word got around fast about a talent show. Bands from all over the globe came to audition and to actually meet the wife of Nikola Justice in person. It was three rounds of judging and the judges were Alyssa, Chi Chi and Amy. Amy was of the models for Alyssa's clothing line but she knew music. Some of the contestants came to the stage solo and some had a whole band. Over 40,000 people showed up to display their talent but only three would be lucky. Each individual only had two minutes to prove themselves on stage do to the mass of people.

It was so much talent out there and this was going to be a tough decision. Hours went by, it was several groups that Alyssa loved but still no decision was made yet. Chi Chi wasn't a big help she didn't like anyone yet, and Amy had a few she had her eye on. The next band came out, they were from the United Kingdom. They were unique with their introduction they had lights flashing and fireworks shooting into the air coming onto the stage. Alyssa loved the energy coming from this band. Alyssa noticed the band was two guys and a girl, they looked like they were in their late twenties or early thirties. One of the guys was excellent on the drums, the other guy was bass and the girl was lead guitarist and sung. They played a song from the 70s that was popular in London.

Chapter Eighteen:

Alyssa stood up and could feel the energy and passion that the group was putting in, and at that moment she found her band. It took the entire day but I guess the saying is true " save the best for last". After the contest ended Alyssa had a talk with the winners and she was pleased with the performance. Alyssa gave them one of her business card and told them to call her tomorrow, and everyone could go out to eat and get to know another. Alyssa made it home and kissed Nikola and told him about the contest. Nikola told her he loved her and he will support a hundred percent whatever she needs. Chi Chi was tired she went straight to her room, and Nikola waited up all night for Alyssa.

Nikola had something special for Alyssa, he had candles and rose pedals with a red carpet rolled out headed towards the Jacuzzi. They had an entire room with just a humungous Jacuzzi with palm trees and flat screen televisions in the room. The couple relaxed in the warm water and the love they had could never die. The kissing was too passionate and Nikola was embracing her body even more, because it's been hours since he seen her. The romance was starting then Alyssa said wait. Nikola replied, wait for what? Alyssa told him about this exotic novel she came across and she wanted to try some new romantic positions. Nikola asked was she serious and can it wait.

Nikola was excited for the moment but he would do anything to please his woman. Alyssa wanted some "bedroom boom" Nikola carried her to their master bedroom and laid her across the bed. Nikola went to start the slow jams on his iPod surround sound. They started to kiss and Alyssa laid down pulling Nikola on top of her. She wanted to try the splitting bamboo sex position. That position is more of the guys job and the woman would lift her leg over his shoulder, and he straddles her other thigh and enters using his hands to support her elevated leg to support himself. Alyssa hands are free to show her clitoris a little self love or to stroke Nikola as he moves in and out. Nikola surely kept his woman pleased, he was a master of science. Alyssa loved to talk at romantic times like this and it didn't bother Nikola because he was a talker also.

After two hours they both passed out across the bed Alyssa went to sleep with her head on Nikola chest. The sun shined bright that next morning into their faces but they still remained sleep. Alyssa phone was ringing but after that workout her and Nikola did last night she didn't want to leave bed. She hopped up because she remembered she gave the winners of the contest her card and she wanted to meet with them. She called the missed call back, it was the band from London and she told them to meet her at the Benihana Japanese Restaurant in Los Angeles in about two and a half hours and if they couldn't find it to Google it or give her a call back. Alyssa had taste for some Japanese food and also this restaurant performed a great show slicing the food.

They got up to go shower and get dressed. Nikola was excited because he just purchased a new car and was exited to drive it. Chi Chi was going with them because she would be the manager of the group and handle the paperwork. Nikola goes to his garage and he has a brand new 2016 Chevrolet Camaro ZL1 red with black rally stripes convertible. It was a beautiful machine and it was just the perfect day to let the top back. He loved California because the weather was always beautiful and he could ride and show his love of his life off. It has never been a man more happy to love a woman the way he loves his wife, and the irony of it is Alyssa knows and she feels the same way. Nikola was the only one in the entire world with this year car because the CEO thanked Nikola personally for designing the cyborgs to manufacture the cars; it was a gift to Nikola.

They made it early to the restaurant and the band was already there. Alyssa introduced herself and her husband Nikola and Chi Chi. The girl from the band name was Aubrey and the boys name was Chase and Ashton. Aubrey has a short punk rock hairstyle and Ashton had a Mohawk now. Chase had long blonde dreadlocks. Alyssa surely was happy with the uniqueness the group has. Aubrey wanted to thank Nikola because his teachings made a believer out of everyone. She said "In London no one had faith and always thought so negative, but after Nikola started teaching the laws of the universe people started believing in thyself and a shift in consciousness took place. Consciousness and energy creates reality and we believe in that ever since we heard it." Nikola was so happy to hear that he motivated this young group. Chi Chi asked how old were they, Aubrey said "I am twenty nine years old and will be thirty in February, Chase is thirty one years old and would be thirty two in the summer and Ashton was thirty four and birthday is on Halloween."

It was the middle of March so Aubrey just had her birthday. Nikola asked what the name of the band was. Alyssa had that planned out already and the name was Jay Cee. Everyone loved that name and the group was getting excited just thinking about being in a real live band. Chi Chi was hungry and she told the chef to start the show cutting up the food. It was so much happiness from the band, because they were young still and never imagined being in a band they said thank you to Alyssa all night, they also learned to be grateful and always say thank you. After everyone had finished eating they were walking out the restaurant and Alyssa to Aubrey she was going to call them after she checks and makes sure her Guitar was finished being made so they could start practicing. They were staying at the best five star suites in Los Angeles so they were close.

On the way home Nikola told Alyssa he had to head to Spain for more teachings. She wasn't able to go with him so she had to stay behind this trip. Spain was late on getting Nikolas teachings across the globe but Nikola would do his best, he always has to help humanity. After Nikola made it home he gathered clothes and his suitcases, for his trip to Spain. Alyssa told him about her time she went to Spain and how beautiful it was. Nikola was going to the Rosa Flamenco restaurant to give his lecture. It had five business rooms and a stage where they play music. The next morning Alyssa took Nikola to the airport, his flight was early. He kissed his wife goodbye and told her to have fun with her music and he can't wait to see her on television. Alyssa replied "Nikola to return home to me safe." She watched Nikola off and Alyssa headed home, it was a lot of work to be done.

Alyssa called her friend who owned the guitar center to check the status of her custom made guitar. They told her the engineer who was making her guitar was out of town, he had a death in his family it was urgent. Alyssa was upset but she needed to get a guitar as soon as possible. She asked did they have the 1960 Gibson Les Paul Standard in stock. The worker told her they had only one, Alyssa told them to hold it she was coming to purchase it. The specs on this guitar were different from the previous one that caught her attention. The body of this guitar was built with a maple capped mahogany design with two PAF hum buckers, truly one of the best instruments made. Alyssa could play rock, blues reggae and even jazz with this instrument. Guitar center had the perfect sunburst color, the guitar was beautiful. They asked her could she play and she started playing for them, while they swiped her credit card. This beautiful instrument will last a lifetime to have paid 250,000 thousand dollars for it, just a quarter million.

One thing Nikola taught her is to always be grateful and life is meant to be abundant in all areas. If she seen something she wanted and to always get it, she deserves to treat herself. They took pictures with her in the store because no one ever purchased this design of Gibson guitars. Alyssa told them not to butter her up she has already purchased the one she needed and just waiting on the one she wants. She also added "its too many millionaires in California now, surley one of them will get the other guitars." She continued to tell them her band name and she was really exited to live her dream. She called Aubrey's cell phone and asked them were they up and ready to practice she just now purchased her new guitar, Aubrey and the guys were ready, and Alyssa told them she would pick them up in half an hour.

Chapter Nineteen:

Nikola made it to Madrid, Spain, and he loved that the people of Spain were so friendly to him and the limacine picked him up. The Rosa Flamenco wasn't that far from the airport, so they arrived fairly quickly. Nikola loved the marble floors and the atmosphere was excellent. The Flamenco dancers, the vocalist and guitarist were all superb and they were giving their best performance. The manager of the restaurant seen Nikola and introduced himself to him, his name was Miguel and he walked Nikola to the back where there was enormous size rooms, for lectures or other business matters. It was already filled with Doctors, Scientist, and Physicist because they all heard of Nikolas teachings but now they actually got a chance to meet him in person. Nikola asked the groups, "What happened if you become injured in your brain? If you become injured in your brain you can not think properly. Now what is thinking? Would thinking be consciousness or would it be a process of the brain moving consciousness though it electrically?"

Nikola answer them "Thinking is an emotion, an electrical response, a chemical reaction. Albert Einstein was exactly right about everything is energy! Humans only access one third of their brains and the body is made up of seventy percent water. What holds all this together in the body? The answer would be energy, humans are an energy field." Nikola was living proof that reincarnation was true because he was a 36,000 year old god, present in modern times. He told the audience that energy only changes form you can not create it or destroy it. Your simply here on earth to wear the garment and to also live your experience, there is no black board with rules you live by. God is in everything, it's the will of life and there is no good or bad in God that's impossible. Nikola told them a story that Alyssa encountered in the celestron galaxy. Alyssa was brave and an ordinary person on an average day and rescued six indigo children and returned them safely to their families. An average person, on an average day, with an extraordinary experience.

Everyone stood up and clapped with joy and was so thankful for Nikola. He told them all to govern their thoughts because our thoughts are our feelings. When you visualize you materialize, so put your attention to the things you want out of life and your undivided attention to the things you don't want in life. The universe is always working no matter what day it is. Nikola asked everyone after they go home to start saying the great word of God (Universe), (I am that, I am). Say that out loud or mentally either way the universe is getting your thought frequencies. "My daughter Keri lee left the planet earth and even me as a scientist didn't even notice her powers. She was the most beautiful girl I ever seen and she saved my life. My daughter could access 90% percent of her brain, from birth we were curious about the pink hair growing in her head. I didn't realize it until years later when my daughter asked if someone couldn't access their full brain what would be the reaction. I answered her they would be looked at as a black sheep or an outcast because people fear what they don't understand. No one would except humans flying and traveling dimensions even though I have experienced it in my time. I worked with my daughter daily to peruse knowledge and her body experienced new chemistries and holograms beyond her wildest dreams."

Nikola finished asked were there any questions? There were no questions so he went to go take a break and would return in an hour. It was Jay Cee first practice and the group was pumped up and excited. Aubrey was in love with Alyssa's Gibson Les Paul guitar, and her new goal was to purchase one. Alyssa always had a beautiful voice but it was new to hear her sing, Chase played the drums for ten years and also in high school, Aubrey played bass and guitar since her teenage years and Ashton played the piano and keyboard in college six years ago. Alyssa was lead guitarist and lead singer, and Chi Chi was in the recording booth ready to record. Alyssa had written a song and it was her passion to sing it, the name of the song was Cosmic Love.

The song itself was about the man she loves Nikola from outer space came and just swept her off her feet. They been to the moon and back and her love for him is deeper than the ocean. After Alyssa told the band about the song they were even more excited, and were ready to play. Amplifiers and speakers were all set up in the garage of the Justice residence and the song was a love rock type song with jazz spin to it, but the voice of Alyssa made the song. Alyssa had so much energy playing the guitar and singing, they nicknamed her Bird. She was flying left to right and was a professional on the guitar. The neighbors didn't complain at all because the song was so beautiful it didn't disturb them. The song put joy in ones soul, it was unconditional love that channeled in tunes. Alyssa was born in the 1980s so she loved the music from that era. Three hours went by fast for the session and it was time for a break. The neighbors walked over and were so happy for Alyssa. Alyssa told her that Joan Jett, Debbie Deb, Bananarma and a few other bands inspired her.

She always loved music just never was able to live her dreams until she met the man of her dreams to motivate her. Jay Cee was well on there way to the big screen, but it was much more work to be done. Alyssa new a few music producers and she asked Chi Chi could she get their CD over to them. Chi Chi had recorded the Cosmic Love song and even though it was practice Alyssa was thrilled that it turned out excellent. Alyssa goes in the house to grab some bottles of water, and then she glanced at family picture of her, Nikola and Keri lee. It brought back so many memories and she wondered where Keri lee was but she knew that she was alive. For some reason Alyssa could feel Keri lee's energy ever since she left the planet earth. Chi Chi came into the house and told Alyssa that the email was sent and they are waiting on a response from the music producer. She noticed Alyssa looking at the family picture and she told her, not to worry Keri lee will be home soon. Nikola was more energized after the break because his passion is to teach and this audience loved him. One of the scientist asked Nikola a question because he did not understand when the soul leaves the body is it like an out of body experience? Nikola answered him "Yes! Leaving the body is just like an out of body experience. Like I said earlier energy only changes form but it can not be destroyed. An individual is considered to be a heavy three-dimensional object vibrating to the hertz of the planet, which allows you to have the same stability of mass itself. So if your field was to suddenly change then the mass that you were made up of would change also, meaning you would be vibrating of the world but not in the world because you are no longer obeying the laws of gravity and physics."

Chapter Twenty

It was a profound lecture given by Nikola, it wasn't a moral lecture but Nikola was expanding ignorance into knowingness. Nikola left headed to his five star suite, they had it prepared for him. Spain is one of the most beautiful countries Nikola has been to, but he didn't care much for the bull fighting and Spain was known for that. He decided to take a walk and do some sight seeing and he noticed people looking at him smiling and because he was putting out great wavelengths that they could catch. Nikola was so positive and confident he did not need body guards because he told himself if he thinks that someone would hurt him then he knew he would manifest it. He approach the riverbank and seen a very large yacht, but it wasn't a regular yacht this was a mobile shopping outlet. It was a grocery store, casino, shopping mall, and art museum all in one.

Nikola read the sign where it circled the river bank every two hours, so he would not get stuck on a boat for hours. He entered and the art was so beautiful, it was all types of bull fighters from decades ago and he seen a beautiful Mona Lisa picture. It was all types of tourist on the yacht it was a perfect time for sight seeing. Nikola didn't bring is camera to take pictures but his cell phone camera was just as great. It was a men's clothing store next door to the art museum and Nikola just wanted to browse. Nikola entered the store and a woman greeting him with a smile and gave him a sombrero, Nikola smile and told her thank you. He put it on his head and continued walking to browse the clothes. The ride of the boat was so smooth he couldn't even tell he was moving. Nikola didn't see anything that caught his attention so he continued out the door to the next store.

The center of the yacht was gorgeous and it was marble floors with a giant flag of Spain engraved into the marble floors. Nikola headed towards the center then he noticed a gorgeous black inharmonic piano, this masterpiece had his attention for moments. It was a gathering here and it was though it was a show or something but the guy in the tuxedo was asking random people, could anyone play the inharmonic piano? No one could play or even answered him and then he asked Nikola. He answered the guy with YES! This was one of Nikola's favorite instruments, he use to play the inharmonic piano decades ago. He asked Nikola what he knows about playing the inharmonic piano. Nikola told him "This is a piano divided in two, and a sharp note and a flat note are not the same." The people were impressed and Nikola sat down and was excited to play. People gathered all around him and cheered him on, and more people started to leave the stores and head towards the center where the music was coming from.

Nikola played the piano for about twenty minutes and it was the end of the song. The applause was loud, they were pleased and asked him was he a music artist? Nikola told them he was a scientist and he was here in Spain, giving his teachings. Nikola started explain music frequency bands to the audience, and everyone gathered around him. "If you can open your brain to contemplate an outrageous concept the frequency of the music goes further to open it. The music I played was frequencies and did you notice how it may have gone right through you? Well they are supposed to go through you because it is all energy. The brain responses to frequencies, and I know it is random but my passion is science so I wanted to share this information with you."

After giving a small lecture Nikola had more fans and they asked him where he was located to learn more from him. "I will give everyone my website and you can see the links where I do live presentations and past presentations. I also have live instant messaging where you can message me or email me. My number is on the website I love giving knowledge out and enlightening people." Everyone took notes and wrote the information down Nikola gave them. The yacht had circled the riverbank and it was time for Nikola to exit. He headed back to his suite to get some rest. He called Alyssa to see how she was doing and to tell her about his trip. Alyssa was happy to talk to her husband and told him about her progress with the music. Nikola was so excited to hear about her progress and couldn't wait to get back home. He didn't know how long his stay was going to be in Spain but he would check on his wife everyday.

Nikola told Alyssa he loved her and good night he was tired, and Alyssa was just getting up. The time zones were different in Spain and the United States. Nikola slowly was going to sleep then started dreaming back to his home planet Nibiru. Osho always told Nikola "Pursue knowledge and as you travel the universe to not persuade different entities but to bring the knowledge to their attention." Nikola would discover in each galaxy there would be different belief systems of thought. His father told him "The only reason entities have addictions to their beliefs is for someway out of a mistake, but does that diminish the greatness of God (Universe), No it does not, it actually opens the coral and lets God be eternal.

The law giving to us here on Nibiru are the same laws giving to the people in the Nebula galaxy, but it doesn't end the universe is forever. Take care of your brain and body Nikola, the brain is the king, the body is its kingdom. Your dreams are like a thunderstorm in our brains and the electrical strikes creates a chemical cascade in our bodies for conquest. My son you are young and wise Nikola and have much to learn still, but remember always use your great intelligence to help others and never for power. We are gods but we also are energy beings like everything in the universe. You never had a chance to meet your mother oh she was the most beautiful woman I ever laid eyes upon. She wasn't from our galaxy but love has no limits my son always remember that. Your mother was a............ (Scientists barges in the room) Osho, Lord Uran is on his rampage again. We must destroy him Osho!" Osho tells the scientist "No, we do not destroy unless there is no other way out. We shall banish him to the planet Mars. The frontal lobe is the crowning achievement in any entities brain, and it allows the individual a distinguishes to change ones mind. Maybe Lord Uran will change his evil ways, think about well all changed from lesser minds to the greater minds we have today. Nikola you stay here, we have to handle this." Osho was a leader on planet Nibiru and he always said that leaders don't create followers, leaders create more leaders.

After failing to enslave the civilization on planet Nibiru they were sending Lord Uran to planet Mars. Mars was closer to the sun and the heat wouldn't destroy Lord Uran because in the past there were great civilizations that build space columns on Mars. Osho didn't believe in war so he always had faith in each individual, and believed in change. Lord Uran was a brilliant scientist and could serve great power, if he was to change. They scientist had Lord Uran inside one of their spaceships to send him to Mars, but Lord Uran wouldn't say one word to them. Osho set the spaceship coordinates to head towards Mars and right after the spaceship took off, explosives went off on the planet. Lord Uran had planted explosives and poisonous gases through out the colony.

Osho was the first to get hit with the blast, he was badly injured. Nikola was inside but he felt the shakes of the blast and looked out the window. He headed out the door to exit and his father was right outside the door bleeding. Nikola was in shock and crying, his father told him to remain calm there is no death. He told Nikola to live the experience of life, and to pursue knowledge. Go meet a beautiful woman and make a family and love them unicondionally. Nikolas father told him to leave the planet Nibiru and spread enlightenment through out the galaxy. Nikola didn't want to leave his father, he never met his mother and his father was the closest relative he had. He was a young boy and he tried to pick his dad up but he couldn't.

Osho told him to go now, and Nikola finally left. Nikola went to the emergency docks and could escape the planet until the gases and explosives were over.

Chapter Twenty-One:

Nikola cried inside the spaceship, the explosives started heading towards him and then he blasted off. He didn't have a destination he just cried and didn't care where he was headed. Nikola had awakened out of his dream, and was in a cold sweat and breathing fast. All he could think about is his daughter Keri lee. He tried for weeks to keep the pain inside and to not think about his daughter's absence but she was his only child and as a father he smiled on the outside but was sad inside. He cried for hours after he awoke from his dream. He knew Keri lee was safe but he wanted her to return home. Nikola got out of bed to go wash his face in the washroom, it was morning time. The sun shined brightly into his suite, it was a beautiful day. Nikola decided to go visit the Temple of Debod in Madrid City but the temple was actually a pyramid under the water in the middle of the city.

The temple was originally built 15km south of Aswan 2,200 years ago. The temple was the most beautiful at sunset but Nikola was interested in what was inside the pyramid. He quickly got dressed and he still remembered the song in his head that was played at the Flamenco. The song was Farruca Solo Compas and the guitar was perfect on this song. This motivated him and he was out the door, and took a taxi to the center of Madrid City. There were small pyramids above the payment but underground was the giant pyramid. It was a stairway downward to enter the pyramid. It was for tourist and this pyramid was surely a beautiful place if anyone was visiting. Nikola seen all the ancient statues of Horus and ISIS it was amazing. ISIS was the Egyptian goddess of feminine wisdom. Nikola had seen all types of translations on the walls the further he walked through the corridor, and the further he went the darker it got.

The lights were dim in this area because the writings on the wall wouldn't show with light. It was the energy of the pyramid, but the average person couldn't read these ancient writings. Nikola could read it with no problem and was aware of most of what the ancients spoke, he seen where the writings stated every entity created was born with a sun and moon zodiac. The Egyptians even had the ancient Miyan calendar on the walls. Nikola knew that the calendar only meant then end of Aeon (age) and into a new constellation of the zodiacs. The pyramid went miles and miles ahead and Nikola continued his search. As Nikola continued to move forward he noticed the writings, started to change into symbols. Nikola didn't really understand the alien like symbolism but he would not give up. Nikola went to get a torch, and tried to see if some light may holds some answers. Nikola held the torch close to the walls, and still no results. The images on the ceiling were revealing now that he brought the torch.

Nikola started looking at the weird coding, but after a few minutes he figured it out. It was thirteen wealthy families over the globe of the earth, and the objective for the next few decades is to try to reduce the global population from seven billion down to seven million. Nikola already knew that the foods were piousness and the vaccines shots but he didn't know what more the elite could do. He continued looking and he read where the elite would try to make humanity destroy each other with guns and violence. Years from now they would try to pull operations to remove gun control from the people. Nikola was never big on war but guns can provide protection and he said to himself if guns kill people then pencils misspell words, cars drive drunk and spoons make people fat. He didn't want humanity to be helpless against a terror like this.

He knew this wasn't the plans of anyone human this had to be a higher intelligence. The ancients were brilliant they put everything in writings so it will always be remembered. Nikola had to find out who was planning this and he didn't know what to do, humanity was so peaceful now that he has changed things. But he also knew it will always be good and evil you can never have one without the other. The next writing had a war going on but it looks like alien armies coming out of the ground. Nikola didn't understand if they were aliens how could they come out of the ground without being seen on radar. Then he had a thought to himself, they only way that was possible is if they were already here underground on the planet. If they take away guns and have humanity weakened from medicines and foods they will have no chance.

Nikola knew whatever this sinister plan was, it would cause deaths. Its going to be bad over the next few years it's going to be deaths from guns, different diseases and panic. Nikola continued to walk after reading these translations but it was dark still and he didn't notice the stone that was up in the ground just before he tripped over it. The stone was sticking up from the ground, the pyramid was old and it was not all the way in the ground. Nikola hit is head really hard on the stones. After about thirty minutes a man passed him and noticed him lying unconscious on the ground. He was an older man so he went to go get help to lift Nikola. The old man found help so they took Nikola back to the surface. They didn't know who Nikola was, so they checked his pockets but all he had was the key to the suite he was staying in and his cell phone.

The gentleman was polite he didn't leave Nikola he took him to his house because he was closer to that and the injury on Nikolas head was pretty severe. They laid him across the bed, and went to go get warm towels and bandages to try to stop the bleeding. They had to stop the bleeding because the hospital was on the other side of town. The young man had to get back to the pyramid because he had his family visiting on vacation there. The old man walked him to the door and told him thank you for the help, after he left he checked on Nikola. Nikola started to wake up, and his head was throbbing, he asked the man were was he? The man smiled and said "You are at my home and my name was Napoleon Libbit." Napoleon was a great quantum physicist and was about six feet in height. He had a very long beard he has been growing for decades.

Napoleon asked him, what was his name? Nikola from that point on started to act different, he told Napoleon he didn't know his name. Nikola was suffering from amnesia. "You are in Madrid, Spain and from the looks of it you're not from here." He told Nikola it looks like your going to have to stay here until you get your memory back. Nikola looked around and see all the experiments and asked what the water experiments on the table were. Napoleon told him "That is a small water experiment I am doing with my thoughts. I took a sample of water dropped it at about 5 cubic centimeters and put it into fifty trays. I took the trays and froze them in a freezer at minus 25 degrees Celsius for about four hours. Then take the frozen samples into a refrigerator that is set at minus five degrees Celsius where a microscope with a camera is set up." Nikola didn't even know what he was talking about so he stopped him, in the middle of a sentence.

Napoleon asked Nikola was he hungry? Nikola said he was starving, and he never ate Gambas Al Ajill (grilled shrimp in garlic sauce). He walked around Napoleons house and looked at all the beautiful art he had. Napoleon was into poetry and had a profound amount of paintings by Leonardo De Vinci. The food was done and Napoleon had set the table and it was time to eat. Nikola was starved he started with one piece of shrimp then grabbed another. Napoleon laughed at how Nikola was eating so fast. Nikola phone started to ring, he didn't even know he had a phone. He looked at the beautiful woman which was Alyssa on his caller ID. He answered with Hello, and Alyssa asked why haven't he called her today and checked in she was worried. Nikola said "Sorry but, who are you?" Alyssa thought he was actually playing, but she was so excited about getting her record deal. She told him that the music producer made some moves and got her in touch with the right people to get her signed for a record deal.

Nikola whispered to Napoleon and said "This woman must think she knows me," Napoleon only talked Spanish so he couldn't talk to Alyssa. Napoleon told Nikola to tell her that you hit your head and that you have amnesia and maybe she could tell you who you are. Nikola started to ask Alyssa but about three seconds later his cell phone went dead. Napoleon knew he was Nikolas only hope in getting his memory back. He remembered Nikola has his key to the suite he was staying in, so they would just go and get the charger. The area Napoleon lived in was a few hours away from the suite but they would need to take the bus or subway train because he didn't have a car. They made it to the underground subway station and the train would arrive in about thirty minutes.

Nikola told Napoleon "Thank you again for saving me." Napoleon said "No problem, it was just doing the right thing." The train came ahead of time and they got aboard. The train would get them there in twenty minutes because it was the new high speed bullet train. He asked Nikola could he remember anything, but it was all a blurry picture to Nikola. Inside the trains were set up like the inside of airplanes so the seats headrest had seven inch LCD flat screens inside them. A Revlon lip stick commercial had came on and it was woman with dark hair and green eyes and Nikola pointed to it. Napoleon asked him did he know the woman on the screen. Nikola replied she looks like someone he may know. The train attendant walked around and asked did they need anything to drink? Napoleon was flirty, he was and old guy who loved the ladies. They both had their attention on the attendant, and then another commercial came it was Alyssa's (Girls Go Green) clothing. Nikola got a glimpse of it and then looked away.

Chapter Twenty-Two:

The train made it to their destination and they walked off the train. Napoleon seen the suite and they headed to it. Nikolas suite was on the 5th floor so they didn't want to take the stairs instead they headed towards the elevator. Suite 505 was to the left of the elevator but after they entered they noticed the room has been cleaned. The maids had cleaned his room, so they took the elevator back to front desk to ask where his bags were. Nikola told them that he was staying in suite 505 and he was looking for his belongings. They looked into the computer for last night staying and he seen Nikola did stay there. The clerk said "Thank You for staying with us Mr. Justice and here is your bags." They looked through his bags but he did not see his phone charger or his identification. He asked the clerk where his stuff was. The clerk called the maids center and asked did they see a phone charger and Mr. Justice Identification? The maids told them we brought down everything we seen.

They told the clerk thank you for the assistance and headed towards the lobby to think. Nikola sat down on the couch in the lobby and Napoleon was admiring a painting on the wall. It was a painting of the great Nikola Tesla. Napoleon told Nikola that the man Tesla was a genius. Nikola looked at the painting and memories started to hit him as he looked at the Tesla painting. Nikola had got up off the couch after he heard his name because it sounded familiar to him. Napoleon told Nikola that we will get your memory back and to lets head to the park to clear our minds. It was a small park just west of the suites. They sat down and started feeding the crows and other birds. Napoleon noticed the birds were all around Nikola as he was the birds' owner. Nikola laughed and started to pet and feed the birds. His memory was gone but his pure heart and wavelengths wasn't. Napoleon has never seen a more peaceful man in his life. A man to have lost his memory and has nothing but still has a heart of gold. A man and woman were strolling through the past and just so happen to pass Nikola.

The guy said hello Mr. Justice. Nikola said hello and did he know him? The guy smiled and told his wife that Nikola was amazing. He said "I am the guy in the tuxedo from the yacht the other day." The man told his wife that Nikola played the piano like no other. The man just stopped and wanted to be polite and speak. Napoleon got up and asked the man did he know Nikola? The man said "The world should know this man, he is brilliant. This man played a piano and gave a speech after that telling us about the music frequency. We checked out his website that he gave us, and this man has done so much for humanity and that's his purpose here in Spain to teach it here." Napoleon told the man how Nikola lost his memory and he was helping him regain it. "We went to the suite he was staying in to retrieve his cell phone charger but apparently it is missing along with his identification." Spain didn't have Nikola's type of phone, he was the only one with it and the only secondary charger he has was at his home in California.

The man and woman started to walk away but before he left he told Napoleon that he seen on television that a person with a memory lost can easily regain his memory with the most powerful human emotion. LOVE! After saying that, he and his wife left and Napoleon had to figure out what to do next. Napoleon remembered what that guy walking in the park with his wife said. If Nikola was a scientist then he would love a science museum. They headed to the science museum maybe it would unlock some memories in Nikolas head. The science museum was huge it contained everything you could imagine that science could define. Nikola looked around and started to notice the planetary sculptures they had. It was an area with planets, galaxies and moons hanging from the ceiling. Nikola loved this section; he read it description of the planets.

He stopped and told Napoleon he feels something is missing on the outside of Pluto. The museum only contains the nine planets in the solar system but Nikola knew something was missing. What was actually missing was his home planet Nibiru the tenth planet in the solar system. Then Nikola looked at the planet Saturn description it said that Saturn has thirty moons and Nikola said this is wrong, Saturn has sixty-two plus moons. Napoleon was exited because he was learning something new and Nikola was regaining is memory. Nikola went to a different area, which had the zodiac constellations. He seen the Leo constellation and went to the Gemini constellation next, but he didn't find anything. Nikola and Napoleon went to the last section which was different universes and galaxies. When Nikola looked at the different galaxies he thought about the rose universe and how the law of the universe applies to us all.

Napoleon asked Nikola, "What is it?" Nikola was regaining more of his memory. Nikola said "The frontal lobe is the crowning achievement of the human brain." Napoleon asked him, "Who are you?" Nikola replied "I AM A AMBITIOUS GOD!" Nikola told Napoleon "Thank you for helping me regain my memory." Nikola reached for phone and seen that it was dead. Napoleon told him that they went to the suite where he was staying and his Identification along with his phone charger was missing. Nikola had a smirk on his face and opened the back of his phone and reversed the battery. His battery on the opposite side ran off of solar power so he didn't need a charger. Napoleon was speechless to see this new technology Nikola had. He had a bar code on his Identification card so he could track it from his location.

Chapter Twenty-Three:

Back in California Alyssa and the band was getting prepared to do there first concert. This was a big event and Chase was very nervous but Alyssa told him to man up. Aubrey had changed up her hair she wanted a more funky look. She decided to get a Mohawk and colored it red and it was rocking. Alyssa wanted a different look also, and her nose being pierced with a small diamond was new to her. She also had her hair down longer with a bang covering her eyes and had on one of her Girl Go Green dresses, it was a grey cotton dress with a red belt and the most lavish heels. Her high heels were the same color red as her belt but the heels had five red buckles going down. They left out the dressing room and headed down the long hall to the stage. The closer they got the louder the crowed got and the more nerves, anticipation, and excitement rose. The announcers introduced the band and the crown got even louder. Their Cosmic Love song was a hit smash and had gotten around pretty fast. Even though it was extremely hot in the arena everyone loved the band.

It was beautiful because the audience had not seen a woman sing and play the guitar like this since the 1980s. Alyssa was getting excited seeing the crowd cheering them on, she danced over to Aubrey and they both were playing the guitar and singing on the same microphone. It was thousands of people even still in their cars from the parking lot cheering for the band. The song ended and Alyssa was out of breath but she thanked the audience from the bottom of her heart for coming to the concert. Everyone screamed "WE LOVE YOU ALYSSA!" They left the stage and went to the dressing rooms behind the stage. The band was so exited they did their first real concert and it was amazing. Alyssa said that she never imagined being a music artist and living her dreams.

Alyssa told the universe thank you and also thanks for friends in the band. She always had an attitude of gratitude no matter where she went. She called Nikola to let him know how the concert went. Nikola answered and he was so happy to talk to her, she was even more excited to tell him all about the concert. Nikola told her he knew she could do it. She told him she loved him and had to go and she would call him back later. Alyssa told the band they did great and she was going home to shower, she wanted to change clothes. After she arrived at home, Chi Chi said "You guys did a great job." Chi Chi had seen the live concert on television. Alyssa went upstairs to go get in the hot tub. She ran a bubble bath to relax and wash the day away.

Alyssa was in her own world now and life was abundant to her. In the back of her mind she missed her daughter again and wanted her family together again. She got out of the bath and went to go get dressed, she decided to go look at some new furniture at the stores she felt it was time to home decorate again. Alyssa asked "Chi Chi do you want to ride and get out the house for some fresh air?" Chi Chi was more than happy to go for the ride. They left in Nikolas convertible Camaro and it was a perfect sunny day for the top back. Most of the furniture stores in the area had the same old furniture from the previous week so she wanted to go to a new store. Alyssa turned the music up and put her sunglasses on. Three guys in a Dodge Charger pulled up beside Alyssa at the red light. The guys started to tail spin showing off the Hemi engine in the charger.

Chi Chi said "I think they want to race." Alyssa didn't even pay the guys any attention, but the guys yelled at her saying when the light turns green then we race. Chi Chi put her seat belt on and Alyssa just smiled. The light hit green and then Alyssa shot off like a rocket. Those guys in the charger were still at the light and Alyssa was twelve seconds ahead and she was laughing at them because the ZL1 engine was too much for those guys. She finally made it to a brand new furniture store just built in Los Angeles. They saw fish tank beds, fish tank toilets and actually a fish tank sink. They were speechless to these unique items the headboard to the bed was not a actually headboard but it was a giant fish tank. Alyssa just needed some new pictures and a sectional for the first floor in their home. She walked around the store, didn't really see much that caught her attention. She seen a few accessories for the washroom so she wanted those and seen a bathtub shaped like the ying and yang symbol that really intrigued her.

Alyssa had very good taste and took pride in things she did. Alyssa didn't see a sectional she was interested in so she just purchased the washroom accessories and left. Driving through the city of Los Angeles in a futuristic car and being the lead singer of a band really drew lots of attention to Alyssa. Fans where ever she went waved and smiled with joy. Alyssa called up Aubrey to see if she wanted to grab something to eat, Aubrey was more than happy to eat. Alyssa asked her could she be ready in an hour and then she could pick her up. Nikola's restaurant No Gravity had opened up across the nation. It was the same restaurant, club, comedy set up just like in Miami, Florida. Aubrey was amazed to see a building so humungous with a club and restaurant inside. Aubrey felt so at home, because Nikola hired people who would love to work and make others happy.

The employees loved their job as much as the pay rate Nikola started them at. The women who worked there saw Alyssa and were so happy to seat her. Aubrey told Alyssa "Thanks again for giving me a chance because I never thought I would ever get this far." Alyssa was happy to hear she thanked her and said "Life is meant to be abundant." "You are a Divine person Alyssa." Aubrey said she liked the planetary atmosphere she was in. Nikola had marble on the floors with giant planets on the inside of the marble. Alyssa educated her on the planets and the laws of the universe. Aubrey was really intrigued and didn't know the planets played a profound role in humanity. Aubrey wanted to move to the food area that had the comedy stage on it. She likes to watch entertainment while she eats.

Aubrey was amazed that the restaurant had iPads for the ordering menu. After they ate Alyssa wanted to give Aubrey a tour of the club. Alyssa showed her each different level had four different clubs. Aubrey loved the indoor beach with sand and the climate control temperature. All the clubs was designed to lead to the next club, because Nikola wanted everyone to feel comfortable with the next person. They went to each club from techno, to rap, to a rock n roll but the techno club was different. It was mainly for dancing but it was so unique to Aubrey because on the dance floor the words appeared as a hologram and read the words of the song. The ladies dances and had a few drinks. It was well deserved because they worked hard for it.

Alyssa called Chase and Ashton to come up to the club and join them. The guys were excited to come. The boys were already close to there and stopped by. They were also intrigued when they arrived and seen the club. Chase asked could he smoke marijuana in the building. Alyssa said "You can smoke in here besides the club has smoke vaporizers that will diminish the smoke in seconds." He asked did Alyssa smoke. Alyssa told him she didn't smoke, but the beauty about marijuana it's a gorgeous plant from the earth and it's actually good for the human body. A famous rapper from California smokes large amounts of it and is IQ level went up dramatically. The DJ started to play Cosmic Love but he put his DJ mix to it and made it a dance song. The crowd got even wilder and they went onto the dance floor to dance. Disco lights were bright, and they were having a great time. The walls were LED lights that also showed the waves patterns of the beats to the music.

The CEO of Pacific records was also in the club. He told Alyssa he admired her club and wanted to say that she did a great job. He was flirty with her and asked her out to dinner one day but Alyssa told him she was happily married. The CEO told her "Marriage means nothing these days." Alyssa looked at the CEO with a smile and said, "That statement is very true, marriage mean little but the heart means everything and my heart is with Nikola, and if you will excuse me I am enjoying the night with my friends. Have a nice time in our club Mr. CEO. Time flies when you are having fun!" It was late and Alyssa was getting tired. The weather was still great driving home Alyssa just was fantasying to herself how great her life is.

Years ago she was working in a department store pay check to pay check not thinking she was good enough. She had her diploma from high school and just thought life was working and stressing day to day. But the man she loves gave her the motivation. With that motivation she became the most loving mother and the happiest woman alive. She always thanks the universe daily, for everything she does. Life is what you make it to be, that is a fact but she also discovered it is also an experience. Alyssa understood why none of her past relationships ever worked. She knew the only way to love was without condition. Things of true virtue you cannot find on the human body as openly as blue eyes or dark hair, true elements of character are hidden behind scheme. Alyssa finally made it home and Chi Chi was passed out on the couch sleep, so she went quietly up the staircase.

Chapter Twenty-four:

Ever since Alyssa went to the Celestron Galaxy she hasn't been having dreams about Keri lee that much. Alyssa wasn't tired now so she decided to stay up and write in her journal and start designing new clothing for her line.

Dear Keri Lee,

How are you? Are you safe? Are you healthy? Are you happy? Are you...there? It has been two years since you left earth and everyday you are not here it feels like my heart will stop beating any moment. We have not stopped looking for you but no matter where we go, your not there and a little piece of me dies slowly. I have been having the same dream every night where I find you and run to hug you but I always fall threw you as if you were a mirage and I just lay there and cry. No matter what I do I do not stop thinking about you in hopes that I walk into the house and your there. We still have your birthday balloon that you loved so much that your dad made special for you with your picture on it saying "Happy birthday Keri Lee, It's your day". I look at the family picture on the wall and pray to the universe that you will return to us soon and complete this broken family. I know that your are thinking about us because I keep feeling this intense aura that feels like yours and even Chi Chi tells me that you will return home as soon as you feel ready but that is starting to feel like an eternity. I started a band and dedicated our first hit to the two loves of my life you and your farther Oh how I wish you could hear it and experience this with me. Nothing will stop me from finding you somewhere in the universes and bringing you home.........NOTHING!

It wasn't a competition for Alyssa unlike the other CEO's of companies. Competition puts you behind the ladder. Other designers were jealous of (Girl Go Green) because Alyssa had the number one entrepreneur on the charts as a woman. She actually was in the top five entrepreneurs in the world but it wasn't the fame or the money which made her happy, because she didn't have a poor soul like most of the millionaires Alyssa found love and peace even before the wealth. She lives a 200 % life. One hundred percent spiritually and one hundred percent material she understood knowing thyself. One of her designs on shirts were inspiring quotes, like if someone is obese to still feel wonderful and fantastic thoughts control the human body.

Alyssa had girlfriends who always say all men are dogs or never trust a man, just all negative things and Alyssa always told them that they're creating the reality they live. "If you tell yourself something on a constant basis then you're allowing your mind to believe it. It is not just about being positive, it's about FEELING positive also. There is no right or wrong in life it's only an experience. All people are divine people." Chi Chi came upstairs and asked Alyssa did she have any aspirin because she had a bad headache? "Try smelling one of those green apples downstairs. The apple scent relaxed the muscles in the brain, instead of you taking medication." said Alyssa. It was morning so Alyssa decided to go do some yard work before it got extremely hot outside. The yard was so beautiful, it was two huge palm trees in the front yard and around the entire home were Orchids, Lilies, Roses, and Sunflowers. The backyard was huge because they had the corner house so they had more than enough room to plant.

Alyssa went to go water the plants because it would get hot later in the morning and the plant would need it to survive. She is truly a woman of nature and in love with the earth. Plants also are a living organism just like the earth. The sun started to heat up fast in California so she went back in. It was a library in their home so it wasn't really necessary for them leave if they needed to read in peace or search for a book. Alyssa would write about everything in her journal to Keri lee so one day when she returns she can read everything see missed. Alyssa wrote down how she has the smartest best dad in the entire galaxy and that her mother loves her dearly. She feels Keri lee close to her all times like she is near or in her presence. It was strange to Alyssa but she didn't turn away from stranger because she has seen the galaxies, helped other life forms, and never dreamed of that possibility. She went out to the sun deck with her journal to think and write.

Day twelve, year two

I think about my first daughter everyday I take a breath. It is nothing like seeing your kids happy and an actual family. I know everything happens for a reason but what is that reason? Family is really important and we all should love one another and build a connection with one another. Yes technology has made us separate and lazy so we don't have time to go out and have a face-to-face connection with someone. I still remember the day you were born and you were supposed to be a little boy. You surprised your father the first day and me. You were so unique in the hospital bed with your black and pink hair and gorgeous eyes. Your father writes in this journal also, he is away out of the country now in Spain. We seen our daughter thirteen years, and she left us. We love her and still wait for the day she return. You're the daughter of the best man I know so you have his knowledge and the heart of your mother. I hope whatever your doing in outer space your inspiring other just like your father and me. But if your not doing that then live your experience. Well I wrote enough in my journal today, please know we love you and come home safe

love, mom.

Alyssa was really emotional for the moment, but she kept a positive attitude. Stress causes ulcers and other health problems but with a positive mind and feeling positive the human body will not get ill. She still went ahead and watched some of Keri lee's videos they recorded of her growing up and her birthday parties. Chi Chi even walked into the room and cried herself. Keri Lee was seriously missed but they all knew she would return. Alyssa had to stay busy so she asked Chi Chi did she want to take a flight out of the country. Chi Chi replied "Yes!" Alyssa checked on the Internet for some rates to travel. She wanted to go to Hawaii. The airlines had some great rates, same day travel matter of fact. It took Alyssa about four hours to get her bags ready and she and Chi Chi left for the airport. She called Nikola to see did he want to take a plane and meet them there but he must have been sleep because there was no answer so she left him a voicemail.

Chapter Twenty-five:

They arrived at the airport and headed straight to the plane. After a few hours they arrived. The people in Hawaii greeted them directly off the plane, and were so friendly. Some of them heard of Alyssa from her band and asked for autographs. They headed to the suite to put their bags up and the suite was beautiful. It was a huge two room suite, when you first walk in there is a living room area with magnificent statues symbolizing love and peace and a sectional made of black leather in front of a huge 73' television. Behind the sectional was an enclosed deck that consists of a hot tub looking out the big window that looked over the whole beach. That afternoon they headed to the beach, she needed a get away because she works so hard.

They changed into bathing suits and went to get a sex on the beach drink. Alyssa was watching her weight so she had a sugar free drink. It was so many different things to do in Hawaii they even had tanks for people who wanted to go diving. It was groups of seven who could go with one instructor so Alyssa and Chi Chi both wanted to explore the ocean. They suited up and headed towards the water. They both had been under the ocean before back in France after they escaped the government chase. Now it is hands on and Hawaii has volcanoes under water. Deep in the ocean it was so beautiful and they saw different sea creatures and schools of fish. There was an octopus that was huge but it didn't bother them and Dolphins swam around them, but they were all friendly.

The entire group had underwater wireless headgear so they could talk to each other. The instructor gave them a tour towards a ship. Chi Chi made a joke asking was it the titanic, it wasn't the titanic but it was huge like it. The instructor told them it was a battleship that sunk from the waves of the Bermuda triangle. Alyssa was wishing Nikola was here to see this. As they continued to swim they saw a network of blocks, a network of blocks that build an underwater city and on top of it was a giant pyramid that reached from the ocean bottom to the surface. It was amazing! Chi Chi told Alyssa "This underwater city you see before us was once the land of our ancestor the natives. After the earth experiences a global shift decades ago it caused floods on the earth."

The instructor was given directions to not to take tourist that far down, but he was even more curious himself than the tourist. They noticed the blocks were moving like it was a door opening, but it was blurry because they were underwater. They all decided to go back to the surface until a group of beings who came from the city below and blasted nets at the divers which caused everyone to start to panic. Everyone scattered but the kidnappers were too fast and caught everyone in the group, one of the tourist drowned because during the panic his oxygen mask had gotten removed. Alyssa opened her eyes and noticed that it wasn't humans that kidnapped them...... they looked like lizard men. The lizard men took them to their underwater city and threw them into small room, similar than jail cells.

The group was scared but Alyssa stood strong along with Chi Chi because they were more focused on being at the bottom of the ocean with little air left in the tanks. The lizard men came back and took the prisoners to another area where the leader of the lizard men was. None of them could believe their eyes they actually had dinosaurs still underground. It was oxygen in the city building so they were able to breathe but the lizard men took their tanks and smashed them. The leader talked in a weird language and no one could understand it. The reptiles started talking in English so everyone could understand it and the leader introduced himself his name was Deserius. The instructor asked him, what did they want with him? The lizard king told them "You were being too nosey and in our domain. We exist underwater by no one knowing we exist.

We have many lizards in your human society everyday. We can look reptilian and change into humans. We have reptiles all over the globe, the news, politics, and entertainment, everywhere. We spy on humanity through our network of satellites, routers CCTV cameras and more and we know who the woman is with the creature, your Alyssa Justice and married to the great Nikola Justice that lives in California. We haven't tracked you in years because apparently you went out of space and we can not track your wavelengths anymore. Nikola can't save you, his technology doesn't even compare to ours. We noticed your husband, has changed things around the globe. He has enlightened millions on the great laws of the universe which we kept hidden to feed off humanity ignorance. We allowed Nikola to spread his knowledge because humans are unstable. No matter how much wealth they will have you creatures always fight. You judge another by race you kill another over material shoes and clothing.

We honestly laugh and tell earthling jokes about your kind. We have different species of reptiles but we never judge another. Where did you humans go wrong? Why not kill you, before you kill each other. We need water to survive because water is a big conductor of energy and I am sure with Nikola being a scientist he has taught you the water experiment with your thoughts so do you now see how powerful water is. You humans poison water, destroy and have no respect for another. Lord Uran was a god and his ignorance got him killed. He wanted to enslave every planet he went to which was completely ignorant. We lizards only want the planet to ourselves. You small minded human let us do as we please instead of opening your eyes to observe the population rising but not in human rising but us lizards populate. You humans kill each other off and argue over music, politics, religious beliefs and much more. You can give a human a good football game they won't even notice life passing them by. Lord Uran corrupted you all with his technology and poisoning your foods and programming your brains through the media and that's fine we don't give a shit. We have a code that whatever another god does we won't interfere with it.

Now we had already prepared our strike against earth, because sadly it is more of us lizards than humanity. We laugh everyday watching your media and you believe what you been told. I think Nikola is the smartest out of you all but no one can save you or stop us. Humanity will stay decreasing and on their knees begging for life. We populated decades ago in east Africa gold mines. We remained hidden for years, undetected. The instructed told the lizard king he was a dam lie about the human population decreasing." The lizard king pulled up his database to show him the population. "We also control the weather and can cause earthquakes and hurricanes and still blame it on Mother Nature and humanity believes it. We have to thank Lord Uran for making humanity not question anything because the sodas and foods he has it sets off a chemical reaction in your small minds to keep you inside the paradigms not to believe anything. If you're told something everyday people will believe it." Alyssa asked, "What are your intentions with us?" King Deserius replied "I have no plans but you all will die, and even the funny looking creature will suffer but the irony of it we wont kill you unless we have to, we will keep you six here and let you kill each other like you humans already do."

Chapter Twenty-Six

Alyssa looked around and seen it was a baby on a table that was being used as an experiment. King Deserius told her "Yes we even experiment on you worthless human children creatures because one day a human may could rises and be intelligent enough to stop us. Kids could be born with all of their brains active and not just 10%. We experiment, punish and eat humans. Take the prisoners away to the, let them rot in a cell. Oh we don't have human food down here so ether you die from thirst and starvation or you can be like us and eat human flesh." It was just six people left after the one person drowned being kidnapped. Chi Chi was upset and didn't know what to do but she knew they had to do something. The instructor and the other people were upset with Nikola and tried to blame him.

"This is what the lizards want us to do fight each other, don't fight we need to save our strength," said Chi Chi. It was dead bodies in the cell from prisoners before them and it was impossible to swim to the surface without oxygen tanks. Everyone was still in shock because of the kidnapping and lizard men beneath the ocean. Alyssa told everyone to be brave and do not give up. Chi Chi could sense Alyssa was nervous about this, but she kept her head up and didn't have any doubts. One of the prisoners said that the lizard men made some important points and facts about humanity we do fight and kill over petty small things. Alyssa told him that is true but we can't focus on that right now, we have to save our strength and get out. King Deserius summoned for one of the humans to do an experiment on. The guards had decided to grab Alyssa but the other prisoners didn't want that. Alyssa quickly told Chi Chi to run out of the jail cell and she could try to find a way to contact help.

The diversion worked because Chi Chi ran past the guards and hide inside a small crack in the wall. The instructor punched one of the lizard guards and they took him away instead of Alyssa. Alyssa and the other people screamed "NOOOOO," because they took him away and they wouldn't see him again. They could hear him still screaming further and further down the hall. Chi Chi was still inside the crack in the wall, she continued further down. She could get a glimpse of the lizard men's laboratory and how they tortured humans. They stung him with scorpion venom and right after inject him with the vaccine to cure him. In the next room they were injecting humans with the DNA of reptiles, changing humans into more of them. Alyssa was talking to one of the prisoners, and got to know her better. Her name was Sade and she was actually born and raised in Hawaii. She always goes deep sea diving but she didn't expect to get kidnapped today.

Sade said "This is not good for me because I'm pregnant." Alyssa replied "YOU'RE PREGNANT!" Sade told her it was dumb to go diving but she got into an argument with her boyfriend and he made her feel worthless so she wanted to do something wreck less. Alyssa told her "Listen love is the strongest human emotion but never for who you are. Never allow anyone to steel your happiness. We have to get you out of her because you stressing would lead you into a miscarriage." Sade was ready to end her life until Alyssa gave her some enlightenment. The guards came back to get another prisoner.

They were headed towards Sade until the last man said take him instead. The guards agreed because Sade was next so they honored his death. He told Alyssa whatever she has planned to move it quickly, so his death wont be in vein. Alyssa started to cry because she didn't know what to do. Chi Chi came back and seen the guards taking another prisoner away. Alyssa asked her did she see any means to escaping. Chi Chi told her she seen no way out and Alyssa told Chi Chi we have to get out of here Sade is pregnant. Chi Chi said "I wish Nikola or Keri lee was here." Alyssa cut her off and said "That's it! Chi Chi you're a genius! If I can govern my thoughts only on Brzee I could guide it here." Chi Chi said "That's brilliant but the only problem if you barge in with the spaceship it would flood the entire city below."

"Exactly, I never tried to bond with the ship though, only Nikola has mastered that. I did help Nikola repair the ship after Lord Uran destroyed it." Sade asked what Brzee was. Alyssa answered her it's a long story and she would explain to her later. "If this works Chi Chi Brzee would arrive in about two hours from California to Hawaii and its worth a try." Chi Chi said "You're the wife of the smartest most intelligent man ever created, I have faith in you Alyssa." Alyssa started to mediate in a yoga posture style. She put her attention only on the spaceship and nothing else. She could feel the ships energy all the way in California. Brzee was activated and was reading Alyssa's thoughts and it flew out of their garage and headed directly towards Hawaii. Everyone was really irritated because of hunger, but help was on its way. The guards came to get another person to torture before they are turned into a reptile. One of the reptiles noticed Chi Chi was missing, but the other one said she was useless because she was not human.

They headed towards Sade, they reached for her then Alyssa stood in front of her so that they would take her away instead. Sade started to cry but Alyssa told her she promised she will come back for her. "Chi Chi will look out for you until I come back. Please trust me have faith I will return." They took Alyssa away and Chi Chi was about to attack them but Alyssa knotted her head no she didn't want them to hurt Chi Chi. Alyssa told the guards they wouldn't get away with this, and to rethink what they are doing. She was really just stalling time with them until Brzee came. Alyssa started resisting from the guards trying to take the key from one of to release Sade and Chi Chi was looking from around the corner at the same time. The clumsy guard actually dropped the key from her resisting getting on the table. Chi Chi moved quickly to pursue the keys because when the spaceship comes it is going to flood the entire city.

One of the guards called for backup, because Alyssa was giving them a hard time she was not going down without a fight. Chi Chi reached the cell and unlocked it. As they were leaving guards were headed towards them on the way to help out with Alyssa. They seen Chi Chi but she didn't want to fight with Sade there, she couldn't risk her getting hurt. They started to run the opposite way towards where Alyssa was then they opened fired at Chi Chi but they made it around the corner out of the line of fire. They sounded the alarm that the prisoners were on the loose, and the King Deserius was not happy with that. He told the guards not to kill them because they had no where to run to. "Capture them and put their bodies on ice but depose of the small pest with them." Chi Chi and Sade made it to the area where Alyssa was but the guards surrounded them, giving the ladies no room to escape.

Chapter Twenty-Seven:

King Deserius arrived and told the guards to feed the feline to the dinosaurs. The captain of the control room radioed the lizard king and told him it was a moving object coming at high speed. He asked was it a missile? No it wasn't a missile, but it looks like no one is driving it. "It is in range for firing Sir, should we take it?" King Deserius said "Yes open fire!" The spaceship took hits but did no damage and bust right through their defenses. This impact allowed tons of water to enter behind it. Brzee went directly towards Alyssa, with tons of water following it. The alarms sounded because the city was about to be flooded. The alarms unlocked all the dinosaurs from the giant rooms in which they were trapped. The lizards were still trying to capture Alyssa even though the city was about to be flooded. The lizard king headed to his submarine air craft for a quick escape.

The reptiles could swim and breathe underwater so it wouldn't be bad for them. Alyssa could feel Brzee getting closer, she told her friends on her count make a run towards the door. She began to count "One, two, three, GOOO!" They started to run towards the door, and everything starts to shake. You could feel the vibration of the tons of water coming. The spaceship came directly towards Alyssa, and the guards started to open fire until the water flooded the entire room. Alyssa put the jets on full blast and headed towards Sade and Chi Chi. She was about eight seconds ahead of the water so the girls had to jump into the spaceship quickly. Alyssa made it to them and said get in. Chi Chi jumped right in and Sade right behind her. The glass top closed on the spaceship and the water immediately came over it. The current of the water was so great the ship lost control but Alyssa tried to take back control of the ship. She started talking to Brzee, saying "Come on girl, you can do it."

The lizard men were being slung around by the current like rag dolls. The engine started quickly after Brzee heard Alyssa's lovely voice. It flew through the water with ease. The entire city was underwater in minutes, and the lizard king had managed to escape, they found an exit and headed straight to the surface. The city started to rise after they left. With the city raising it was causing a typhoon sucking everything in the area into it. They continued to the surface even through the typhoon. They made it to the beach shore and landed. That was an adventure for Sade she never experienced anything like that, it was breath taking. Chi Chi told her she was use to it. Alyssa told Sade to live a peacefully life and make sure she has a healthy baby and don't stress it only does damage to the body. Sade listened to everything Alyssa told her and left to for home.

Alyssa and Chi Chi were just lying in the sand, taking a break from all of the action then they heard a beeping noise it was coming from Brzee. Alyssa had got up and walked over to check it out. By Alyssa controlling the spaceships from her thoughts it actually triggered something in Brzee. Alyssa could understand what Brzee was saying but it was not in a language but more in a sound vibration. She asked Chi Chi could she understand the noise. She didn't even hear anything she thought Alyssa was just hearing things. The screen inside the ship came on and Osho appeared. Chi Chi told Alyssa "That is Nikolas father on the screen.

"HELLO WHOEVER YOU MAY BE, IM GLAD MY SON NIKOLA HAS FOUND LOVE AND PEACE IN HIS LIFE. IF YOU ARE WONDERING WHO I AM... I AM OSHO.... NIKOLAS FATHER. I BUILD BRZEE, NAMING HER AFTER MY LOVELY WIFE THAT I LOVED. NIKOLA NEVER MET HIS MOTHER. I DESIGNED BRZEE WITH ARTIFICIAL INTELLIGENCE, TO OPERATE THROUGH THE LAWS OF THE UNIVERSE (LAW OF ATTRACTION). NIKOLAS THOUGHS CONTROLS IT, BUT TO UNLOCK HER FULL POTIENTIAL IS WITH THE POWER OF LOVE. BRZEE HAS BEEN WITH NIKOLA FOR DECADES AND HAS PROTECTED HIM WELL BUT NOW YOU HAVE AWOKEN HER AND SHE IS VERY MUCH ALIVE BUT ONLY TO THE ONE WHO LOVES ITS MASTER NIKOLA. IF YOU TOUCH THE LOWER PANEL OF THE SCREEN A PICTURE WILL APEAR. SHE SHALL SEE A FAMILY OF FOUR. IT WAS ME, ON THE LEFT, MY WIFE BEHIND ME AND THE LOWER LEFT IS NIKOLA AND THE RIGHT IS NATHAN. NATHAN IS NIKOLAS TWIN BROTHER. (Alyssa and Chi Chi eyes got three times as bigger after hearing that). YOU MAY NOTICE THE PICTURE IS BURNED AND RIP, THAT WOULD BE BECAUSE OF THE EXPLOTION THAT TOOK PLACE ON OUR HOME PLANET NUBIRU. THE TWINS WERE SEPERATED AT BIRTH NATHAN IS OLDER THAN NIKOLA BY ELEVEN SECONDS. NIKOLA HAS THE INTELLIGENCE LIKE HIS FATHER AND NATHAN IS CONTROLLING AND MISUNDERSTOOD LIKE HIS MOTHER. NATHAN WILL SENSE BRZEE POWER THAT HAS BEEN AWAKENED BECAUSE IT'S THE SAME ENERGY AS HIS MOTHER. NIKOLA WILL LEARN IN TIME HIS MOTHERS' HISTORY AND HE WILL MEET HIS TWIN BROTHER. PLEASE TAKE CARE OF MY SON FOR ME, LOVE HIM LIKE A MOTHER WOULD LOVE THEIR OWN CHILD. HE NEVER EXPERIENCED WHAT IT WAS TO HAVE A MOTHER. SINCERELY OSHO JUSTICE.

Chi Chi was speechless because she has been with Nikola for decades and never knew he had a twin brother, we learn something new everyday. Nikola must see this recording he will be amazed to know he had a family. His memory must have been wiped to protect him. They decided to head back to the suite to get some rest but Chi Chi didn't go to bed because it was not late. Alyssa was drained from today's adventure and after she took a shower she laid across the bed and immediately went to sleep. She started to dream, but this time the dream had Nikola in it.

They both were in the Nephilim galaxy. (The Nephilim galaxy was consisted of entities that were half demon and half angel.) When Nikola seen Alyssa he hugged and kissed her as though it was real. After a while they both realized that they were actually alive in the dream. They continued on and noticed they were walking on pink frozen water crystals. He noticed that the large frozen water crystals were projections of someone's thoughts. Nikola was many things but stupid was not one of them he knew his science. He knew what they were because he has experimented on the frozen water experiment many times. It got even more beautiful the further they got because the crystals were in the air now. He mastered the water experiment so he could tell what the water molecules were without even using an inferred microscope. He told Alyssa whomever thoughts these were coming was highly intelligent.

Alyssa asked him, "How did you know?" He ignored her and continued looking because it was very interesting to him. Up ahead was a cliff so they had to stop but the view was even more beautiful looking down from the cliff. It was a gorgeous city but it also had the moons inside the planet in the sky. That was mind blowing to Nikola he didn't know how that was possible. A group of men and women approached them from the rear they asked no questions they only told Nikola and Alyssa to come with them. They went along with them without any resistance. They made it below into the city and the strangers took them to CAPTAINS chamber. The captain entered and asked Nikola what was there business here and whom are they working for? Nikola said they were with no one and wasn't looking to fight. The captain took her helmet off, and her hair was the prettiest pink hair. After removing her helmet Alyssa walked to her and said "Keri lee is that you?" The captain didn't know what she meant what so ever, she told Alyssa she must have her mistaken for someone else. Alyssa started to cry and said Nikola "Its Keri lee." It was actually Keri lee twenty years in the future then Alyssa showed Ker lee a picture of her when she was a baby. Keri lee was shocked looking at the picture but she was still in denial. Nikola couldn't believe he was looking at his daughter in her mid thirties then Keri lee told them to follow her.

Chapter Twenty-Eight:

They all started talking and Keri lee told him what galaxy they were in and apparently it was ahead of their time. It was light years into the future even ahead of the technology on Nikolas home planet. The beings in this galaxy could fly actually and could access one hundred percent of their brains. This was Nikolas intention for the planet earth for humanity to access all of their brains. Knowledge is just the seed to reality! Keri lee told them "Yes it is true, the entities are brilliant here but what does that mean? There will always be people who want power. You can never have good with the bad, negative without the positive." Nikola finally realized that the reason they were in this dimension was because they both had to be dreaming at the same time if they were to wake up out of the dream they would leave from this galaxy. It was strange to Nikola but he was excited to see his daughter.

Alyssa asked Nikola did he think Keri lee was just as powerful now just like back on earth. "I'm not sure!" They both noticed something different about their daughter, like she was happy. A man entered the room he then approached Keri lee and kissed on her forehead, which made her day. Then it dawned upon them they realized that she was in love. Keri lee introduced Vergil to her parents from the past. Alyssa was so happy and started to cry with joy. Even Nikola started to shed a tear, because they never thought to see their daughter grown up so fast and finding romance. Keri lee told them that they have been married for three years now. Nikola asked "Outside of the cliff where we entered this galaxy I seen pink frozen water crystal, those where yours thoughts right?" "Yes, those are my thoughts. It actually keeps enemies out, but my mother here as you call her I could sense good in you both. We have so many different entities who try to attack us here and any given time. My father always taught me to use my power for good and to protect those I love. The people here may be super smart and half demon half angel but still need to be protected. Enemies come here for the water, because out of the four elements water is the most receptive. When you first entered this galaxy I'm sure you noticed the pink lily flowers and that you could walk on water. This is a place of beauty and peace."

Alyssa put her head on Nikolas shoulder and smiled she was so happy to see such love. Keri lee told Vergil to stay with her until the end and Vergil told her not until the end, Always. They both started to kiss then Keri lee invited her parents to a gorgeous ball dance that was tonight. "I could take mom to go get a dress and my husband will take you dad to get a tuxedo. Besides we all can get to know another because this is still overwhelming for me to believe." Alyssa and Keri lee were on their way to the outside outlets to purchase their dresses. Alyssa asked her did she not remember them honestly. Keri lee said, "I remember bits and peaces from my past, but not much as a child. I left earth because I didn't think dad wanted me because he said I wouldn't fit in with other children. I am accepted here in this galaxy not because I am stronger than the Nephilim but because I am myself." Alyssa looked at Keri lee and said "We missed you so much Keri lee, your father and me have been worried sick. We stayed busy most of the time to try to hide the worries we had. We knew you were coming back but we didn't know when. I don't ever want to wake up from this dream because in reality it's a nightmare not having you there. I and your father are quite wealthy but wealth doesn't always define a person. Wealth just brings us freedom to do the things we want to do without any questions." "Your right mother, I learned so much from you and father. Im sorry I left but I want to discover this power I have. The entities here on this planet are wise but no one has ever came along and ever given them sufficient intelligent knowledge about their beautiful self until I arrived. They didn't know how their bodies work from the inside out, why they have addictions."

"Keri lee you have gotten wiser, I am so proud of you as a mother as a best friend. "It is so different here mother, versus in your Milky Way galaxy." Alyssa asked Keri lee was she not coming back to earth. Keri lee said "Honestly mother I don't know! I love it here I met a husband and it is peace, I feel like its paradise. I saved the people here many times over and I feel like it is my job to protect them, but I also know I have parents that I truly love. Like you and dad always taught me it's no right or wrong I am only here to live the experience and wear the garment." "You have grown up so much Keri lee and even more beautiful. I can tell you are so happy but I still want to hear these lost fifth teen years, where you been and what happened after Lord Uran, your going to have to fill me in another time but for now I am glad to be with my daughter. My only question is will this be real after your father and I awake? Will you only be in my dreams?"

"Im not sure mother, I don't have all the answers but we will never know until it happens. Mom you know what? I didn't like that western concept in America that everything must be perfect or people were looking for physical beauty. That's a travesty to the human experience. I wish you both could move here, and how is Chi Chi?" Alyssa answered "She has been fine, just worried about you." "We have to find a cute dress mom, I am actually excited we are shopping together." Alyssa said "Im curious Keri lee where did you meet this husband of yours?" Keri lee smiled and said "Mom he is a nice guy. After I left earth something from the rose galaxy was calling me, it's a long story mom I will fill you in another day. He is a real nice guy and I care for him very much. Mom did you gain weight? Im not calling you fat...... but you look thicker." Alyssa looked at Keri lee and smiled "You were always like your father with a sense of humor, but no your mother still has it. That reminds me, I started a band." "Like a music band mother?" asked Keri lee "Yes Keri lee! We are awesome I love our band, the name is Jaycee. They are the nicest people I ever met, from London. I hope you can meet them someday. We searched everywhere for you, you know? I experienced other dimensions but it felt as though you were there. I have always felt you spiritually."

They went to a small boutique that had the most beautiful dresses Alyssa ever saw. Alyssa saw a red and white long silk dress that flairs out at the bottom with cut off sleeves and a diamond orchid in the middle of the dress and Ker lee found a long light blue silk dress with lace inside and flairs out at the bottom with a bow in the middle. They were done shopping and headed back to meet with the guys. "We made it back from dress shopping. Do you like it father?" Nikola looked and smile "Yes Keri lee I love it!" Alyssa tells Nikola to get his camera and take some pictures. Nikola tuxedo looks very nice it was an all white tuxedo with a grey under vest and a white tie. Nikola grabs his camera, "Smile Justice Family."

Chapter Twenty-Nine:

Chi Chi started calling Alyssa name, she was sleep fourteen hours. Alyssa jumps out of her sleep. She starts to look around just before she hops out of bed. She looks around and notices she is not in her dream anymore. Chi Chi asked her "What is it." Her phone starts to ring, and its Nikola. Alyssa answers "Hello, hey honey were you just in my........" Nikola answers before she finishes her sentence with dream. "Yes my love we were dreaming but it was so real. I cried after I woke up this morning because we were next to our daughter. I can't believe this is happening. I didn't want to wake up but room service was banging on the door to clean the suite."

"We were just about to take a picture with your camera Nikola and next thing I know Chi Chi is calling my name saying I slept fourteen hours. I can't believe we were in each other dreams, it was so real Nikola. The irony about dreams you can never remember how they started. Let's talk about this at home, so when are you leaving Spain?"

Nikola said "Im leaving Spain now I been so busy working but I need time with my wife. We will meet at home tomorrow. You are in Hawaii so you will get there faster. Im glad you are ok by the way. I know Brzee has been touched I got a notification to my phone." Alyssa smiled and said "You make me laugh Nikola, oh yes its something important you must see. I don't want to tell you because it is just my job to spoil it but as soon as you get to California I will tell you. Were are about to gather out luggage and head home. We will arrive in a few hours with Brzee, and she will be fine with no scratches Nikola." "Okay my love see you soon!" said Nikola. Alyssa yelled from the room to the living room "Chi Chi lets gather our stuff we headed home in an hour." Chi Chi said "I am curious Alyssa, what did you dream? Did you and Nikola see Keri lee?" "Yes Chi Chi! It was so weird you know, like she was married and all grown up and Im not sure was it even real. Nikola also was sleeping himself and he was there in the exact same dream. It was called the Nephilim galaxy or something like that, it was a place half angel and half demon the beings were and Keri lee just like her father feels she has to protect them. I didn't mean to sleep all day Chi Chi sorry but Im sure you have had good dreams that you didn't want to wake up from. She had this sword and we walked on water, oh yes Chi Chi her hair was all pink and longer. Remember last time we seen her, she had black and pink hair but this time it was all pink and longer. I wish I knew if what we saw was real." Chi Chi I wish you could have been there, it was amazing. I didn't bring my journal did I this would be the perfect time to write my experience down, but I can wait until we get home. It was just so real words just would not do it justice.

Wow Chi Chi we been talking for two and a half hours its time to head home." They finally made it home and unloaded their luggage and Alyssa looked at Chi Chi "Home sweet home Chi Chi! Do you know we have not spent a good week at home in a long time? We have done concerts, vacations, explored different dimensions I mean Chi Chi I never imagined this, I never imaged my daughter missing. Chi Chi its just................ (Alyssa starts crying). Chi Chi wiped Alyssa's tears and said "Don't cry Alyssa I promise you will see her again, I miss Keri lee just as much as you do." The door opens up and it was Nikola home from Spain, "Hey honey I'm home! Chi Chi why is Alyssa crying?" Chi Chi said "Nikola she is just going through her emotions right now. Women do that time-to-time, we have been though so much and back you know."

Nikola ran over to console her and said "Alyssa don't cry I am home now, and I'm not going anywhere else for a while. I been trying so hard to please humanity and I haven't noticed we been distant. I'm never leaving you again Alyssa or Chi Chi. I have three ladies that I love more than anything the universe can manifest for me. Alyssa look at me, I want you to know I am serious about this. I love you so dam much woman I think about you everyday. Not a day goes by I don't think about you, even when we are together I think about you. Alyssa baby it is no other woman that

I'm in love with other than you. I love without condition your heart is pure and I sensed that the first day we met years ago in France. Now stop this crying, I can't stand to see the woman I love in pain. I'm never leaving again I swear and we will see our daughter again. You are like my best friend, my soul mate you're the mother I never had."

Alyssa stopped Nikola in the middle of his statement, "Oh my goodness! Nikola you reminded me of something, come I must show you. Chi Chi and I went beneath the ocean in Hawaii but that's not important read this message on the screen inside Brzee. "It's my father, Alyssa what happened?" Wow my dad was a genius and I have a twin brother named Nathan? Amazing! Alyssa I always knew you were my world but I never would imagine that you would unlock this. My father said I will know the history of my mother soon enough but I never knew about my twin brother amazing. This is so much to take on at once. My family has been split all my life I never met my mother or brother. My daughter is somewhere in the endless universe." Alyssa grabbed his face "Nikola! Listen I'm your wife don't worry, we will get through this together okay? I don't know the story behind your mother, but I can promise you this we will find out together."

"Alyssa you been like a mother to me, the mother that I never had. Chi Chi can tell you my research has always kept me busy and now in this lifetime I found love. I remember years ago when I had nothing but myself, Brzee and Chi Chi in the trunk which I didn't know, but I didn't have nothing. I always dreamt big and had faith in the universe." said Nikola. "Nikola I know you had faith that is what brought us together just like faith will unite our family back together. You changed the world Nikola, people are wealthier than they ever been. It's not all about the money but we also encourage people to have a pure heart. I remember in the Celestron galaxy I had to enlighten some friends… You know what I think all emotions are connected, like love may be connected to hate and vise versa. Look at what you done Nikola, humanity has never been like this. Which there will always be issues with people but that's a part of everyone living their experience in life. You know the day should be the sunrise and the beginning of great thought. That's something I learned in Hawaii." Nikola smiled and said "Tell me about your trip in Hawaii. Did you two enjoy it?"

Alyssa looked over at Chi Chi and they both laughed, "Lets me say this and I'm done, I will never look at another reptile the same again." "Looks like you and Chi Chi had the time of your lives. Sweet heart you make me laugh so much, I never want you to take your love away. You want me to tell you about my experience in Spain well no matter my love I will tell you anyway. I apparently hit me head pretty hard and got amnesia. A man named Napoleon was so friendly he helped me through that nightmare and I told him thank you a million times. The irony of it is that I seen so many things that reminded me of you, and with that I regained my memory. The power of love is the strongest emotion ever. Amnesia is so awkward because I couldn't remember a thing but if I seen something that reminded me of you and Keri lee I could picture you in my head.

Spain also has a yacht that has outlets inside it, and you can shop and get a tour of Spain it was so amazing. I'm happy to be home with my wife though because we been so busy with our careers we been forgetting family and all we have is us.

How is the band by the way?" Alyssa had so much excitement telling Nikola what they have been doing, "Oh yes Nikola I'm glad you said something about it, let me grab the CD you can hear it yourself its called Cosmic Love, it's our single." As Nikola played the song on their home stereo system he couldn't help but to see how much real talent his wife has, "Yes I love it, and you sound the same as you talk. You know how most British or European artist voices changes completely on television and singing but your voice sounds beautiful like nature itself. I'm in love with a musician!"

Alyssa replied "And I'm in love with a scientist! We have to be the oddest couple ever. You're a scientist from another planet and I'm a European rock star. The beauty of it is love is like energy it has no form, shape or color it just is." "We should never put our dreams aside as imagination, because every dream is the next step of our evolution. Is it not amazing Alyssa, how the mind works. I want to tell the world welcome to the kingdom of heaven (earth), come without judgment, without hate, without testing or without anything. We control the power of intangibility we pull out of inertness action chaos and hold it into which we call MATTER. It's so random I know Alyssa but when we start talking about quantum physics I get carried away."

Chapter Thirty:

Nathan Justice tells his crew that their heading to earth and to set the coordinates. One of the men in his crew named Eric asked "Sir, why are we heading for earth?" Nathan just continued to stare out of the window of his ship, then he answered "My twin brother Nikola, our mom told me about him but I never met him before, he should be happy to meet me because I could feel our mothers energy from earth so I know his aura is pure. Nikola wants to educate and fit in with the humans but I want to show humanity a new way. Humans are like children and don't know any better and are too weak to protect themselves in the current state. I don't want to enslave humanity no I simply want to lead them. My mother would want that even though my father favored Nikola more. Nikola didn't have mother growing up, and I didn't have a father coming up so we are very different even though we are twins.

"That's in the past and this is the now." Another crew member walks up to Nathan "Lord Nathan we just arrived in the Rose Galaxy, what is our next objective?" "How many times have I told you not to call me Lord Nathan? Call me Nathan we are all gods I'm just in control of this vesicle." "Sorry sir!" Nathan did not like the crew to call him Lord Nathan because he wanted them to know they are all gods. "It is ok, I'm just thinking to myself about family matters now. My twin brother is an ambitious god, on the planet earth. Our father gave him all the knowledge but not I, I had to steal, kill, and live more of a harder life. All of us beings are Divine beings now lets get our supplies from this galaxy. We need food, fuel, and few other things to help us on our journey. It's going to take four hours to re-supply the entire ship so we have time to explore this galaxy.

Let's find where the warrior shop is I need a new sword, this new age is nothing like the old days with a great sword battle. "Hello shopkeeper I am Nathan and I'm looking for the greatest sword you could possibly own. I want it pure steel, nothing cheap like these swords you have in the front display. Where are your real swords?" The shopkeeper says "Excuse me sir, but these are the only swords we have in this shop." Nathan looked around and said "Now shopkeeper I'm sure you don't want me and my boys to wreck your shop do you? I hear you make the best swords ever to be weld in ones hand." The shopkeeper did not want any trouble so he answered Nathan "Follow me this way!" Nathan had a confident smirk on his face "I thought you might see things my way. Yes, these are the swords I like, the real battle swords a man can die with a good sword. Decades ago shopkeeper I was the best swordsman ever, and this new age concept with guns came in to the universe. Enough about my past I want the blue Japanese katana sword now that will serve me well. Thank you dear shopkeeper I didn't catch your name but I'm sure I wouldn't need it."

The old shopkeeper nodded and said "Good day lad!" Nathan noticed a group of men harassing a woman as he walked passed the alley. He could here the cries of the woman and he has compassion for women because he learned much from his mother. He asked the gentleman was there a problem but the men knew Nathan wasn't from this galaxy so they surrounded him. He noticed the woman had a black eye and her lip was bleeding. The men didn't get one word out after surrounding Nathan before he used his brand new sword. Nathan was more ruthless than Nikola, and didn't think twice about killing. He cut each one of the men leg off and left them lying on the ground then he approached the woman, and told her sorry about that. He knew she was terrified seeing that but Nathan would react first and ask questions later. He didn't stay long because he still had to get more supplies for the ship. The woman followed him and told him thank you. He took his hat off with manners and introduced himself.

"I am Nathan Justice commander of the Enceladus. We came to this galaxy to fuel our ship and get supplies. I don't know you, but I cannot stand to hear a woman's cries plus my mother I might add raised me. My lady, allow me to stop talking and since I didn't catch your name please tell me." The woman hesitated but then she told him, "My name is Blue haven but you can call Rose." "Rose you say? Indeed I love it my dear Rose was my mother's middle name. My mother meant the universe to me and may I asked you why was those guys bothering you?" Rose answered slowly, "My father is an astrologist and he is onto something big with the universe, something he calls dark matter or dark energy. Seventy percent of the universe is dark energy and I just over hear my dad talking to other scientist about this universe stuff but they try to scare him away from it. I know my father and he is not backing down from it."

"Well I'm sure those guys wont be bothering you again, the pleasure was all mines I should be on my way now, my crew waiting for me." Rose told Nathan, "You're really a nice guy Nathan I wish we could meet again someday and sit down and get to know another." Nathan kissed her hand and said "The feeling is mutual, I will return for you someday to have chitchat with you. Goodbye!" His crew walked up and Eric said "Sir we managed to get food and supplies without you. We couldn't find you so we went ahead and got it. Is everything ok sir?" "Yes captain everything is ok! I rescued a woman and she reminded me of my mother. Let's take flight captain!" "Where to sir?" "Planet earth sounds nice as I said earlier, besides I want to meet this twin brother of mine."

Alyssa was happy to have Nikola back at home and now it was time for them to go out and enjoy each other's company. Chi Chi told them they should do something wild just to get their blood flowing. "What you have in mind Chi Chi?" asked Alyssa. Chi Chi thought about it and said "Honestly I don't know Alyssa but you and Nikola can come up with something I'm sure." Nikola came up with skydiving. Alyssa walked over to Nikola "Yes that would be perfect, skydiving. Nikola looked at Alyssa with discomfort "I was only joking when I said skydiving. Are those parachutes safe?" Alyssa laughed, "You can fly all around the earth and galaxies with no concerns but you're worried about jumping from a plane with parachutes. Yes their safe Nikola now let's go. Chi Chi can come also we need a family day, it has been a while."

They arrived at the sky diving airport just in downtown Los Angeles. It was a group of twelve including Nikola and Alyssa. Chi Chi could fly so she didn't get aboard the plane. Nikola was nervous because this was a new experience for him and he was always use to having his spaceship at great heights like this. It was a small airplane but it would seat about thirty people, so it was plenty of room. The group got dressed into their suits along with their parachutes and loaded up on the plane. Some of the people knew Nikola and told Alyssa they loved her music. The instructor gave everyone directions and if they hand second thoughts speak now. No one had second thoughts, and the plane reached its height. Alyssa and Nikola were the last two to jump because the group wanted to do something special for them. When they all jumped they formed a heart shape in mid air by everyone holding hands. Alyssa kissed Nikola after that and said "That's so sweet of him to set that up."

They both held hands and were jumping on the count of three. Nikola hugged her and started to kiss her, their hearts were so deep into the kissing they jumped at the same time. They were upside down kissing, with love in the air. Nikola looked into her eyes "I love you Alyssa Justice!" "I love you Nikola Justice." Alyssa responded. They couldn't kiss like they wanted to because the goggles were huge on their eyes and Nikola didn't want to knock her goggles off. They fell through the group heart, and they took pictures of the lovebirds kissing. The suits and helmet had cameras on them so every moment was captured. They started to play like kids swinging each other around in the air, and Alyssa went onto Nikola back. He started to pretend he was a super hero and flying her on his back. Alyssa hugged him tighter and tighter she didn't want to let go. They both needed this experience to get the love that was missed back.

Chi Chi flew past them yelling, and told Nikola she didn't know he could fly and they all started laughing. Alyssa was just missing her Keri lee and started to wish she were there. They took pictures of the clouds above, it was a very beautiful scenery. They seen another plane approaching them but Nikola manage to pass it without being hit. Alyssa got off of Nikolas back so he could release the parachute. They both released the parachutes at the same time and Alyssa said that had to be the most romantic experience they ever encountered. "Alyssa you are everything to me, I mean literally we explored the universe and back together." Chi Chi caught up to them and asked how it was? "It was amazing Chi Chi. Without the goggles I'm sure its blinding, I don't know how you could fly all day Chi Chi." They finally made it to the ground and when they returned the suits to the instructor he printed out all the photos they took in the air. It was over one hundred different photos that were taken, so they chose the ones they wanted and then they decided to head home. Alyssa wrote in her journal about her skydiving experience and Nikola went to go ask Alyssa did she want to stargaze into space. Nikola had the best telescope money could buy the light reflected better than the actual Hubble telescope NASA has. The outside deck was the perfect place to stargaze day or night. Nikola actually seen different entities flying in space, but they posed no threat to him. He knew the universe was divided into two parts where Earth is made up of the four elements versus the stars and planets which has a fifth element called Ether. He knew the universe was not static but actually expanding. Nikola was curious because if the universe was expanding it must have emerged from a dense and hot state but he wasn't sure if the Big Band might have caused it.

Gravity started drawing matter together into a web like structures that gave birth to stars and galaxies. Nikola saw a dying star swelling and shedding its outer layers and leaving a small dense sphere about the same size as planet earth. He told Alyssa to look at how dense it was. If two dying stars orbit one another it will draw upon the companion's outer layer. Nikola even had sound on his telescope. It could hear the sounds of the universe. Sometimes it sounded like a shock or a zap sound but he knew the entire universe was electrical. Alyssa looked out at the stars and the beauty of the universe, "Nikola you know what? This just gave me another idea you call this stargazing right? Well I think I know my next hit single for the band. We can rock a song called Stargazing. Let me call the band up, we can get back in the studio and I know Aubrey will love it. I know exactly how I want the stage to be we can have stars and shooting rockets, oh my, it's going to be amazing. Since your back Nikola you can see our performance and practice. I'm so happy your back baby." Nikola kissed Alyssa and said, "My love I'm happy to be home, I thought about you everyday in Spain. I wanted to get you a souvenir but that didn't happen." Alyssa smiled, "Its ok, your all I need in this world. I'm going to call Aubrey and see if she is awake."

Alyssa ran to get the phone and then dialed Aubrey number "Hey girl, it's me Alyssa, are you busy?" Aubrey was glad to hear from Alyssa "No sweets what's up?" "Nikola and I were just browsing the universe on telescopes and I just came up with a fabulous name for a new song. It's called Stargazing! So what you think, do you like it" Aubrey yelled threw the phone "I love the name of that Alyssa wow." Alyssa smiled, "Yes, me too. It just hit me and I already have the stage set out in my head for our first concert. I know it's late I am so sorry to wake you Aubrey." Aubrey laughed, "Oh no girly your ok, I wasn't doing anything just thinking about some new piercings and some epic tattoos." "Ok well call the boys tonight tell them about the new song and I will meet you at the studio tomorrow." "Sounds good will do Alyssa, see you tomorrow."

Alyssa hung up the phone with Aubrey, "Nikola now you will get to see me in action. Baby this is great, I love you." Nikola smiled "I'm glad you found this passion Alyssa, I you love so much. This is going to be awesome seeing my wife on stage at a rock concert." The next morning Alyssa woke up early, "Baby wake up its time for practice, get up so you can watch us." Nikola got up and got himself ready and went into the garage where the band was practicing, "Alyssa you got me up this early in the morning just for your practice?" Alyssa stuck her tongue out at Nikola and said "Yes, I'm excited about bringing my husband to hear me." The band was all here and already setting up for practice when Alyssa and Nikola came in.

"Ok here is the beat that I came up with in my head it has a little bit more enthusiasm in this one and we are going to have to get wild also it's a faster beat. Our producer agreed with me on the beat. Ok when the chorus comes in saying the stargazing part, I can show out my guitar skills and spin the guitar. I've been practicing that at home also I know the perfect outfits we all can wear for this concert." The band practiced for about four hours and Nikola stayed for the whole practice. "We did great today at practice guys see you tomorrow for another great session." A few days later was the concert and they arrived on a gorgeous tour bus. It was a double-decker tour bus that was enclosed at the top. Inside was just like a five star hotel, with four rooms and a nice size kitchen area and the seats were bucket seats just like a lounge in a club.

The band was talking and eating on the bus on the way to the concert, and Aubrey had her tattoo artist ride along also. Aubrey was getting a tiger tattoo in her head, she shaved the sides of her hair to make it visible. Going down her neck was the top of a tree and it continued down her body. It was a giant tree but inside the tree was the story of her life in words as if the tree was speaking. Nikola was well known at NASA so he was able to get some spacesuits for the band to perform on stage in. Alyssa came up with the idea of the band wearing spacesuits on stage for the beginning of the show. Seth started reading the mail that had came in, "Here is some fan letters, saying how much they love us and Aubrey they asked were you single? HA HA HA I love our fans, we have to give them whatever they want." The bus started to slow down and Alyssa looked out the window to see where they were "Ok guys we arrived, wow all these fans outside. There is the beautiful red carpet guys, let's go."

Chapter Thirty-Two:

The crowds' screams Jaycee on and Alyssa smile was bright as ever. They stopped and signed autographs for the fans, they even took a few pictures before they went on stage. Backstage they had interviews with Alyssa and the band. One of the interviewers was from Epic Magazine, "Do I call you just Alyssa or can I call you Bird?" Alyssa told him he could call her Bird. "Ok thank you Bird, but we are with Epic magazine and we just had to interview you and your band. My name is Phil and I'm interviewing live today with Jaycee. Bird you're the number one fashion artist, your married to the world's greatest scientist, you started a band and your on tour right now. How do you feel Bird with all of the accomplishments?" Alyssa smiled and answered, "Well Phil, I feel fantastic or how can I say energetic." "Bird what will you tell the people watching this interview what can they do to live the life you live or be the best rock star of their generation?" Alyssa looked at Nikola and answered, "You must have faith! Believe in yourself and also take action, change is always good. Never let anyone tell you that you can't do it because you can do whatever you want, just put your thoughts to it. Only hang out with people who support you, who believe in you. I started out in a clothing store Phil, I met the man my heartbeats for and he opened me up to my wildest dreams. He gave me faith and with that faith I started everything I have today. It could take a while sometimes but it also takes a bamboo tree five years to mature. Five years Phil, but the irony of it after six weeks it grows ninety feet tall."

"Now was it six weeks or was it five years?" It was five years because without watering it over that five-year period it wouldn't have grown. The universe gives us whatever we desire our thoughts are frequencies we are energy beings and energy is everything." "That was a breathtaking speech Bird, amazing. You and the band have a new hit song tonight right?" The whole band answered, "Yes, we do Phil." Phil smiled "I'm sure Jaycee is going to give us a show tonight." Aubrey answered, "Oh no doubt Phil, we are going to give the fans what they want. We love all our fans and we love the supporters." Alyssa added, "I want to say I love my husband, my daughter, but mainly I want to thank the universe." Phil looked over as Nikola walked to Alyssa, "Nikola Justice decided to join us. Hello Mr. Justice!" Nikola shook his hand, "Hey Phil! I'm just coming to support my wife." Phil said "I never met you in person before Mr. Justice this is an amazing day for me." Nikola smiled and looked at Alyssa, "I came to watch my wife and band rock out I think that's the term." One of the stage crew walked into the room, "It's about time for the show so its time they let you guys go and get ready." Nikola, Alyssa, and the band shook Phil's hand and said "Thank you Phil! Take care."

"That was the Justice family people and I'm Phil interviewer from Epic magazine." Alyssa hugged Nikola, "Thank you honey for coming out to the interview with me." Nikola looked into Alyssa eyes, "Hey you're my wife, anything for you." The stage crew brought the outfits into the dressing room, "Wow these astronaut suits are custom made, I love them. Thank you Nikola! This is going to be one hot show." "These suits are so awesome." Aubrey loved the spacesuits they were performing in. Nikola said, "It was a gift from Alyssa." They were all dressed now and Alyssa wanted to go over what they were about to do "Everyone knows there role right? Seth is going to come down from the ceiling, which the projector will be giving off images to make it seem like he is failing from space. After he comes down, it will be right into the drums. Chase will already be on stage looking through a telescope but he is already on the keyboard. Aubrey and I will have the crowd thinking we are already on stage, but we will actually be coming from the wear on skydiving boards. The theme on stage will show the entire universe, Nikola gave the guy with the projector some images from the Hubble telescope so it is going to be a show tonight. It's Showtime!"

It was millions of fans going wild the concert was completely sold out. The crowd had signs saying Nikola and Alyssa rocks, and we love team Justice. The introduction of Jaycee got the fans even more rowdy. The performance was spectacular, but couldn't expect anything less from this band. It was the passion and the love of the fans that motivated Jaycee. It wasn't a competition or battle against other bands but no other band has ever made any education music about planets or even the universe. It was an excellent song and the words were meaningful. Alyssa was so excited she did something amazing and jumped into the crowd. The fans caught her and carried her around the whole stadium. She didn't stay long in the audience she had a job to do on stage.

The entire building was bright by the planets they even had a giant telescope in the back of the stage that was looking into space during the concert and actually showing what is was observing onto the screen. Nikola cheered her on from the side of the stage. Alyssa looked at him and smiled and blew him a kiss. Nikola was always excited when people were learning and this was an excellent way to get people to understand the universe and it's more than just us out there. Alyssa was looking at the spotlights at the top of the ceiling. She noticed someone was standing there but she couldn't make out who it was until they turned away and walked off. Her mouth dropped because she got a glimpse of someone with pink hair. She stopped in her tracks and looked and pointed up but the audience thought it was part of the show still. She was telling Nikola to look up it was Keri lee.

Nikola didn't hear her but he could read her lips. The crowd cheered Nikola on while he was getting on stage. Nikola looked up and didn't see anything. Later that night, after the concert Alyssa started telling Nikola that she know it was Keri lee she saw. "Are you sure it wasn't just the different color lights?" Alyssa looked at Nikola with disappointment, "It had to be her Nikola I'm sure of it." Nikola starts laughing and Alyssa asked what was he laughing at? "I'm not saying your making it up but honey I didn't see anything and I just don't know." Alyssa stomped off, "How couldn't you not believe your wife? Nikola this is our daughter and you think I'm seeing things, don't you? I can't believe you Nikola you really just pissed me off really. We all have disagreements but I wouldn't dream of making it up with our daughter." Nikola stood in shock "Honey wait, where are you going?" Alyssa stormed out the house, "Don't you dare even look at me now Nikola Justice!" Alyssa leaves home and slams the car door. Chi Chi hears the door slam and ask what was going on. Nikola was standing at the door holding his head, "Chi Chi during the concert Alyssa said she seen Keri lee and I didn't really believe her but I didn't know she was going to blow up like this."

Chapter Thirty-Three:

Chi Chi shook her head at him, "Nikola she is a woman, you sometimes can disagree with women." "I know Chi Chi but." Chi Chi flew up to Nikola face, "No buts Nikola just let her cool off she will come home." Nikola was still standing in the door looking out hoping Alyssa turns around and come back home, "She was pissed Chi Chi I never seen her that upset. Keri lee really needs to come home because Alyssa getting moody." "That is her daughter Nikola!" "I know she is my daughter too, but what do you want me to do? We searched the universe Chi Chi, I even tried to dream the same dream again after going to bed every night. I pictured in my head before bed and tried to dream it but it was weird because I couldn't do it. I miss my daughter just as much as Alyssa. I sit in the parlor looking out to the backyard daily thinking about her or ways how to locate her. Chi Chi if you have any ideas please speak now, because I'm not losing my family. I wonder where Alyssa is headed."

Alyssa went to drive around the city and clear her head, "I give that man everything he wants and he couldn't even believe me. I know what I seen I'm not losing my mind. Or am I? What if I want my daughter so bad, that I am seeing things? I'm not delusional." Alyssa was driving eighty miles per hour in a sixty-five speeding zone. "I am so upset right now I can't even see the road from these tears. I don't care about anything now. I'm losing myself everyday without my daughter. I don't care how wealthy we are we still have family issues and money can't buy happiness. If I do I would be with Keri lee, but she is not dead. I wish I knew what to do. Suicide is not the answer, I am a bad mother. What if she ran away because of me? Nikola always told me not to think negative thoughts but I can't help it. I am sick of this life, I put on a fake smile but on the inside I'm torn. My only daughter has left the earth from me and I don't have the answers. I have no one to talk to but Nikola and he thinks I'm-seeing things. I hate my life and I'm fed up being miserable."

Alyssa was blinded by rage and anger she didn't pay any attention to the truck that ran the light. The truck hit her car and knocked her several feet back. This accident caused a big scene and ambulances' were called and it was on the news. Alyssa was knocked unconscious and was in bad shape. The guy in the truck was telling the police officers sorry he ran the light and he was crying because he thought Alyssa was dead. They rushed her to the closest hospital by med flight Alyssa was not even breathing. Her car was completely totaled. They arrived to the hospital and the nurses and doctors got to Alyssa time she arrived. She was still alive but barley, she was in a coma. One of the nurses recognized Alyssa and told the doctor who she was. They immediately called Nikola.

Nikola was on the beach daydreaming and thinking what he was going to say to Alyssa. His phone was in the car, he didn't hear it ring. Nikola continued to walk through the sand at a loss of words. He kept having flashbacks of Alyssa slamming the door and how he felt bad for laughing at her. Some girls from the beach approached Nikola, and started to flirt with him. They had bikini tops on and were super friendly. One of the girls recognized him and said "Hey you're that scientist nerd guy." Nikola stopped and said "I'm sorry ladies I don't want to be rude but I'm married and I want to be alone now. My wife and I got into an argument." The hospital called the Justice home and Chi Chi answered and they told her the bad news. Chi Chi immediately called Nikola phone and also got the voicemail so she left four voicemails on Nikolas cell phone.

Chi Chi was on her way to the hospital. She wrote a note also and left it at home for Nikola to read once he came back home. Nikola continued to think to himself and was wishing for his mother and fathers guidance. He was out of answers and didn't want to lose his family. Nikola could sense something was wrong with Alyssa but his sadness blinded him for the moment. He started to walk back towards the car to send her a text message but he turned back around and wondered the beach still. He wanted to give her time to cool off plus he didn't have the right words to say to her yet. He seen a flower shop on the beach and he went to go purchase some flowers for her. The flower shop had a large flat screen television, which the news anchor was on showing the bad accident. It was showing Alyssa's car totaled but Nikola was purchasing the flowers during that so he missed it.

The doctors were out of answers for Alyssa she was actually on her own. It was no medication to wake her from her deep sleep. Chi Chi finally arrived and they told her what happened and they asked where Nikola was? Chi Chi told them about the argument they were into and this happened. She didn't know where Nikola was, then Chi Chi told them, "When he finds out he will be devastated, he will come. Nothing would keep him away from Alyssa." Nikola headed back towards his car. He had memorized what he was going to say to her and he had cheered up a little. Chi Chi started crying seeing Alyssa hooked into all those tubes and that big bruise on her head. That impact was really hard. Nikola reached his car and he sat in the driver seat took a deep breath. He had to gather his words for his apology. He grabbed his phone and seen the missed calls and checked the voicemail. He started the car and speeds towards the hospital, was already halfway down the road when he realized he left the door open.

He drove one hundred plus all the way to the hospital. He didn't care if he had police clearance his only mission was Alyssa. Nikola arrived to the hospital parking his car in the emergency room entrance. He ran to the nurse station and asked "Where is my wife? Where is Alyssa Justice?" Three nurses went with him and showed him the way. He busted through the door and fell to his knees seeing her like that. Chi Chi told Nikola the doctors said its nothing they can do. Nikola started talking to Alyssa, telling her to come back to him, he is nothing without her and he couldn't go on without her. "You have to wake up Alyssa I know your soul is still there. Chi Chi we are not giving up she has to come back I know she will I'm positive of it."

Chapter Thirty-Four:

Nikola was there and Alyssa could hear him but she was still unconscious. Alyssa was in another dimension but she thought she was dead. A little girl approached her and asked her why was she giving up? Alyssa was on a beach but still unconscious in reality. She woke up on a beach with the little girl in front of her. Alyssa said "I'm not giving up." The little girl shook her head, "Didn't your husband teach you better? You can't think negative your only going to attract it. You were upset and let your anger blind you. I'm just a kid and you're a grown up you should know better." Alyssa smiled at the little girl, "You're a smart kid, with a smart mouth. Where am I by the way?" "You're in the universe and that's connected to everything silly. Were you trying to kill yourself?" Alyssa answered "Kill myself? No, NO I couldn't do that, or maybe I was. I'm upset ok!"

The little girl stoops down so she can see Alyssa's face, "Well how is that anger working out for you? You are in a deep sleep and it could last for years." Alyssa got up and looked around "This looks like my home state California." The little girl skipped around Alyssa, "Well yes your home but this is the negative side of your current reality. You're lucky to be unconscious. You hit your head pretty bad."

Alyssa rubbed on her head "I was driving wreck less yes I admit it but I lost my daughter ok." The little girl stopped and looked up at Alyssa, "What's your daughter name?" "Keri lee! Her name is Keri lee." The little girl asked, "Does she have black and pink hair?" Alyssa stood is shock, "Yes, yes how do you know?" The little girl answered "Keri lee has been to our dimension before, she didn't think she was loved on your home planet and she wanted to explore the universe. How do you know she has even been gone? She could be right under your nose the whole time but sometimes parents can be so busy with their addictions and forget about the children."

"Yes your so right kid. Wow! I didn't even think about that. So will I ever escape this dimension?" "Well that depends on you and your faith. This is not a test or anything because you know that life is not a test it's an opportunity. This is only your doing your anger allowed this to happen. You can't go driving insane and hoping you don't get into an accident." Alyssa frowned, "My husband and I got into a argument and I was upset because he didn't agree with me. We been married years now and never had one single argument until now. I caused this yes I feel sorry because I know he is worried and it's killing me because I can still feel his presence and hear his voice. It's so crazy because its like he is beside me and I can hear him but I'm not aware. Can you hear him?"

The little girl stood there as if she was trying to listen, "No I'm not a loved one in your family only you can hear the ones you love when you're unconscious. Your mind and heart will find the way for you but you must have the will to do it. It's the will of life that's defines us. I think you're a wise woman Alyssa Justice. This dimension has so many entities just like you who come here and just give up on life. You can never ask a question that you don't already know the answer to. The answer usually is what we don't believe we could do and it interferes with ones agenda with their own addictions that define them. We have never had enough time to even care for another person other than own emotional addictive needs." Alyssa asked the little girl, "Do you know if I will ever see my daughter again?" The little girl laughed and said, "You asked a lot of questions but like I just said you know the answer, you just have to live the experience. I can be complicated at times, but life is what you make it to be. You are not building rockets or anything in that nature, your simply living life and it's a great life."

Alyssa looked at the little girl and smiled, "Ok kid if it's so great why are you here?" "I can leave this dimension whenever I want. I'm simply a visualization of what you were focusing on the most which is your daughter. I'm a girl and a teenager, is that not the last time you seen her on planet earth? I can be whatever you create in your mind. Change your thoughts but you know this already." Alyssa looked out to the ocean "It won't ever happen again, I guess I have been a little moody lately. I feel so bad because my husband is here to support me and I can't even do anything about it. This shows arguments not even worth it, not just because life is short because we are immortals but we only hurt the ones we love. I could just see his face now I know he is crying and I just want to start crying now myself. I'm glad I met you kid but I didn't even get your name."

The little girl smiled and said, "My name is not important because I couldn't be anything, remember you create your own reality you are your own observer."

"Kid you sound just like my husband." Chi Chi noticed Alyssa smiled in her sleep and told Nikola she thinks Alyssa is in another dimension. "Chi Chi your right, how could I be so blind. Of course she is in another dimension." Chi Chi had a confused look on her face, "What do you mean Nikola?" "Chi Chi it's like a dream, or sometimes a nightmare, she is not with us in this reality but she is somewhere else in space and time." "Well will she ever come back to us Nikola?" "That depends on her Chi Chi but I can answer you this, I'm never leaving her side. She will hear my voice in her head so much she will think I'm there." Chi Chi smiled at Nikola, "Nikola you're the best man I know and I have faith she will come back to us. I know you will rest here tonight with her and I can't blame you because I will be right here with you."

Alyssa was still standing there with the little girl, "So kid do you think Keri lee will come home to us?" She look at Alyssa, "Still asking questions are we? I make myself laugh, but yes of course you're her mother and father she has to but first you have to get back to your reality." Alyssa heard something in the distance, "Do you hear that kid? Its Nikolas voice again. I'm getting so happy just to hear my husband's voice. It's amazing how love makes you feel. Its takes over your whole being, and I know in my heart I love that man. I can't believe I'm actually talking to a kid about love. Even in another universe I can feel his love. Can I be killed in this dimension?" She answered, "Yes you can die anywhere, but if you die here there is no going back. I would hope that you don't want to die here seeing how ambitious as you are. Let's go for a ride!" Alyssa asked how. "Easy Alyssa, focus your mind." The little girl started to fly and Alyssa couldn't believe her eyes, "Wow kid you can fly! So can you now concentrate and let's go." "I can't!" yelled the little girl, "Didn't we just talk about how you limit yourself? Stop your doubts and focus." "Okay! Here goes."

Chapter Thirty-five:

Alyssa started to feel herself get lighter and lighter "Wow I'm levitating. This is amazing, my daughter did this." The little girl flew around her, "Well anyone can do it, and you have to concentrate. You see all those stars in the sky? Sometimes I wonder are stars are actually beings who have died or beings that souls are trapped. Some people just give up on life and let their insecurities get the best of them. I see hope in you Alyssa. You just have to want it for yourself because it is pointless for me to believe in you and you don't even believe in yourself. Focus on yourself and you will see your daughter again. You said your husband is a nervous wreck but I'm sure he has figure out by now you're in another dimension. I'm sure he hopes your journey is safe but the danger is not out yet because without faith you can stay trapped here. Those are black holes over there and that's the Milky Way over there." Alyssa looked at the Milky Way galaxy, "Wait I live in the Milky Way galaxy." The little girl answered, "Yes you do but try to head that way it will send you right back. Your mind will get you back but only when you ready. This universe never ends and I love meeting beings with greater or even lesser minds. It's so quiet in this dimension. Well of course everyone here is miserable, lonely, and suicidal, there emotions are not alighted. That's why the frontal lobe in your brain allows you to change your mind at anytime."

"You just had one of those days. How can you say you live fully everyday by simply experiencing the same emotions that we are addicted to everyday? When you have mastered your life you will see the day and have opportunities in time to create avenues of reality and emotions. Your day would become fertilization of infinite tomorrows. Alyssa I sure hope you're paying attention because you may not get a second chance if you do manage to escape this dimension." Nikola was still by Alyssa's side, "Chi Chi I may leave in the few minutes to go get some clothes for Alyssa and me. I remember the first day I met her Chi Chi her eyes shined bright in the store. Chi Chi I wish you could have seen her. We experienced so many different adventures together it was a new experience for her and it made my day. Alyssa has gone to other dimensions she gave me my first daughter Chi Chi. I'm sorry to vent, but she is all I have now. I can't lose her.

Every second she is gone it takes a peace away from me. I hope she is ok. I'm not there to protect her Chi Chi. What if she is taking on an obstacle that she can't handle?" Chi Chi grabbed Nikola by his shoulders, "Nikola calm down Alyssa is family she is tough as nails." Nikola calmed down, "Yes, you're right I'm still overreacting Chi Chi. I am having an emotional breakdown." Chi Chi said, "Nikola now you know better than anyone not to get you're though patterns out of alignment. I know she is your addiction but she will come back to us." Alyssa started shivering, "It's getting colder here now kid is it suppose to be this cold?" The little girl flew beside her, "Oh yes sorry about that, the weather is random here. Is could be tornadoes, hurricanes, typhoons any kind of storm at any moment."

Chi Chi got Nikola's attention, "Nikola look, Alyssa lips are turning blue. She must be in a blizzard. Quick find me some cover that will help very little but it will do something." Alyssa asked the little girl, "What do we do here with this cold weather? We have to get through this. "It's a cabin with a fireplace just up ahead. Don't die on me Alyssa, we have come too far. It's so cold and I can't keep going. Its freezing, I can't even feel my body." The little girl yelled, "We made it! Hurry inside. It's a fireplace already with wood." Alyssa sat by the fireplace to warm up, "I want out of this place. I'm going to scream I can't take this. Tell me what to do. How do I start? I want to escape this place. I'm sorry to cry but this is really frustrating." The little girl sat by Alyssa, "Crying is not going to get you out. Put that energy into focusing."

Alyssa yelled, "Stop giving me these riddles kid, I want out. Just leave me alone!" The little girl stood up, "Your wish is my command. Goodbye!" The little girl disappeared and Alyssa was all-alone in a galaxy she didn't know anything about. "What do I do? I'm all-alone. No one now all because of my frustration. That kid was so nice and just was trying to help me find my way. Ok Alyssa you got yourself into this, you're going to have to get yourself out. Nikola always told me the brain is king the body is the kingdom." Alyssa's lips got their color back and she was warm. Nikola walked over to Chi Chi and said, "Chi Chi I'm going to grab some clothes and over night materials ok. If anything happens please call me. And I mean Anything!" Chi Chi nodded her head, "No problem Nikola!" Nikola walked outside of the room Alyssa was in and started to cry. He didn't want to cry in front of Chi Chi, the nurses came and gave Nikola some tissue for his tears and had told him it was going to be ok. "I'm going home nurse. If anything happens call my cell phone please." The nurse took the number from Nikola, "I will do that Mr. Justice." Nikola walked down the hall thinking, "This can't be happening I can't let my emotions get the best of me because now I am angry. My daughter is gone my wife is in a coma." Nikola left the hospital on the way home. Only thought he had was his wife waking up. It was all on her to wake up, he didn't have the technology to save her with this. Pure love was the only way and the will of Alyssa.

Nikola arrived home and took a hot bath to get his thoughts together. This was a new experience to him so Nikola started to meditate in the bathtub. This was really needed to get Nikola back to reality. He asked the universe for guidance and energy to see him through this. His thoughts were positive now and no more negative thinking. Nikola decided to get out of the bathtub and to go get dressed and gather Alyssa's clothes also. He was back to normal now and he went though the house looking at their family pictures and putting a big smile on his face. This was needed also to keep him in that mood. Alyssa was still meditating trying to focus her mind, and escape this dimension. She couldn't concentrate because she had anger in her heart, which was blocking the path for her to leave. Her faith was weak and she needed some motivation.

Chi Chi was alone with Alyssa and started talking to her. Alyssa smiled in her sleep because in the other dimension she could hear Chi Chi's voice. "We love you Alyssa and Nikola and I are waiting for you to return on the other side. He is being hard on himself now because he is taking the blame for your accident." Alyssa heard Chi Chi's cries and this shifted her even more. Alyssa felt bad for being rude to that kid, she was so nice and helped Alyssa out in this dimension. She felt bad for the way she acted and just kept saying, "I hope she forgives me." Nikola went to go get Alyssa clothes out of their room. When he arrived to Keri lee's room it was dark. He cut the light on just wishing she would have come home safe because her mother needs her right now. Nikola was bout to cut the light off and then he looked on the floor close to the window and noticed a picture. He entered the room and walked towards the picture then he picked the picture up and dropped the bag of clothes he had in his hand for Alyssa. He stood froze for a second. He couldn't believe his eyes. UNBELIEVABLE!!

www.ingramcontent.com/pod-product-compliance
Lightning Source LLC
Chambersburg PA
CBHW051023180526
45172CB00002B/450